A UNIFIED FRAMEWORK
FOR VIDEO SUMMARIZATION,
BROWSING, AND RETRIEVAL

A UNIFIED FRAMEWORK FOR VIDEO SUMMARIZATION, BROWSING, AND RETRIEVAL

with Applications to Consumer and Surveillance Video

Ziyou Xiong,

Regunathan Radhakrishnan,

Ajay Divakaran,

Yong Rui, and

Thomas S. Huang

ELSEVIER

AMSTERDAM • BOSTON • HEIDELBERG • LONDON
NEW YORK • OXFORD • PARIS • SAN DIEGO
SAN FRANCISCO • SINGAPORE • SYDNEY • TOKYO

Academic Press is an imprint of Elsevier

Elsevier Academic Press
30 Corporate Drive, Suite 400, Burlington, MA 01803, USA
525 B Street, Suite 1900, San Diego, California 92101-4495, USA
84 Theobald's Road, London WC1X 8RR, UK

This book is printed on acid-free paper. ⊗

Library of Congress Cataloging-in-Publication Data
A unified framework for video summarization, browsing, and retrieval with
applications to consumer and surveillance video/Ziyou Xiong . . . [et al.].
 p. cm.
 Includes bibliographical references and index.
 ISBN 13: 978-0-12-369387-7 (hardcover : alk. paper)
 ISBN 10: 0-12-369387-X (hardcover : alk. paper)   1. Digital video–Indexes.
2. Video recordings–Indexes. 3. Automatic abstracting. 4. Database management.
5. Image processing–Digital techniques. I. Xiong, Ziyou.
 TK6680.5.U55 2006
 006.3'7–dc22   2005027690

British Library Cataloguing in Publication Data
A catalogue record for this book is available from the British Library

ISBN 13: 978-0-12-369387-7
ISBN 10: 0-12-369387-X

For all information on all Elsevier Academic Press publications
visit our Web site at www.books.elsevier.com

Printed in the United States of America
05 06 07 08 09 10   9 8 7 6 5 4 3 2 1

To my parents Anmei, Xiong and Jinlan Lu, my brothers, Zisong and Zixiang, and my sister, Tian'e.

— Ziyou Xiong

To my parents, Malathi and Radhakrishnan, my sisters, Rup and Krithika, and my professor and friends.

— Regunathan Radhakrishnan

To my daughter, Swathi, my wife, Padma, and my parents, Bharathi and S. Divakaran.

— Ajay Divakaran

To Dongquin and Olivia.

— Yong Rui

To my students: past, present, and future.

— Thomas S. Huang

Contents

List of Figures

List of Tables

Preface

The goal of this book is to develop novel techniques for constructing the video table of contents (ToC), video highlights, and the video index as well as to explain how to integrate them into a unified framework. To achieve these goals, we have introduced a hierarchical representation that includes key frames, shots, groups, and scenes as presented for scripted video and another hierarchical representation that includes play/break, audio-visual markers, highlight candidates, and highlight groups for unscripted video. We also have presented a state-of-the-art content-adaptive representation framework on unsupervised analysis for summarization and browsing. We have shown how this framework can support supervised analysis as well. We have reviewed different video indexing techniques that are based on color, texture, shape, spatial layout, and motion activity. We then have presented a unique and unified framework for video summarization, browsing, and retrieval to support going back and forth between the video ToC and the video index for scripted video, and between video highlights and video index for unscripted video. Furthermore, we have reviewed many of the video summarization, browsing, and retrieval applications that have already used these technologies.

The book is designed for sharing our research results with graduate students and researchers in video analysis and computer vision. We look forward to inquiries, collaborations, criticisms, and suggestions from them.

Acknowledgments

Part of this work (Rui and Huang) was supported by Army Research Labs (ARL) Cooperative Agreement No. DAAL01-96-2-0003 and in part by a Computer Science Education (CSE) Fellowship, College of Engineering, University of Illinois at Urbana-Champaign (UIUC). The authors (Xiong, Radhakrishnan, and Divakaran) would like to thank Dr. Mike Jones of MERL for his help in visual marker detection. They also would like to thank Mr. Kohtaro Asai of Mitsubishi Electric Corporation (MELCO) in Japan for his support and encouragement and Mr. Isao Otsuka of MELCO for his valuable application-oriented comments and suggestions. The authors (Rui and Huang) would like to thank Sean X. Zhou, Atulya Velivelli, and Roy R. Wang for their contributions.

We would also like to thank Mr. Charles B. Glaser of Elsevier for initiating this book project and Rachel Roumeliotis and Rick Adams of Elsevier for supporting the publication of this book.

One author (Xiong) carried out the research reported in the book when he was a Ph.D. student at UIUC. His current employer, the United Technologies Research Center (UTRC), does not warrant or assume any legal liability or responsibility for the accuracy, completeness, or usefulness of any information, apparatus, product, or process reported in this book. The views and opinions of authors expressed in this book do not necessarily state or reflect those of UTRC, and they may not be used for advertising or product endorsement purposes.

Chapter 1 | Introduction

1.1 Introduction

Video content can be accessed by using either a top-down approach or a bottom-up approach [1–4]. The top-down approach (i.e., video browsing) is useful when we need to get the "essence" of the content. The bottom-up approach (i.e., video retrieval) is useful when we know exactly what we are looking for in the content, as shown in Figure 1.1. In video summarization, what essence the summary should capture depends on whether or not the content is scripted. Since scripted content, such as news, drama, and movies, is carefully structured as a sequence of semantic units, one can get its essence by enabling a traversal through representative items from these semantic units. Hence, table of contents (ToC)–based video browsing caters to summarization of scripted content. For instance, a news video composed of a sequence of stories can be summarized or browsed using a key frame representation for each of the shots in a story. However, summarization of unscripted content, such as surveillance and sports, requires a "highlights" extraction framework that only captures remarkable events that constitute the summary.

Considerable progress has been made in multimodal analysis, video representation, summarization, browsing, and retrieval, which are the five fundamental bases for accessing video content. The first three bases focus on meta-data generation and organization, while the last two focus on meta-data consumption. *Multimodal analysis* deals with the *signal processing* part of the video system, including shot boundary detection, key frame extraction, key object detection, audio analysis, and closed-caption analysis. *Video representation* is concerned with the *structure* of the video. Again, it is useful to have different representations for scripted

1

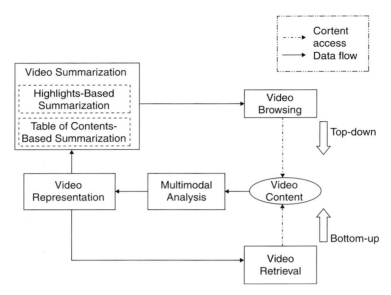

Figure 1.1   Relations between the five research areas.

and unscripted content. An example of a video representation for scripted content is the tree-structured key frame hierarchy [3, 5]. Built on top of the video representation, *video summarization*, either based on ToC generation or highlights extraction, deals with how to use the representation structure to provide the viewers top-down access using the summary for video browsing. Finally, *video retrieval* is concerned with retrieving specific video objects. The relationship between these five bases is illustrated in Figure 1.1.

As Figure 1.1 shows, video browsing and retrieval *directly* support users' access to the video content. For accessing a temporal medium, such as a video clip, summarization, browsing, and retrieval are equally important. As mentioned earlier, browsing enabled through summarization helps a user to quickly grasp the global picture of the data, while retrieval helps a user to find the results of a specific query.

An analogy explains this argument. How does a reader efficiently access the content of a 1000-page book? Without reading the whole book, he can first go to the book's ToC to find which chapters or sections suit his needs. If he has specific questions (queries) in mind, such as finding a term or a key word, he can go to the index at the end of the book and find the corresponding book sections addressing that question. On the other hand, how does a reader efficiently access the content of a 100-page magazine? Without reading the whole magazine, she can either directly go to the featured articles listed on the front page or use the ToC to find which article suits her needs. In short, the book's ToC helps a reader *browse*, and the book's index helps a reader *retrieve*. Similarly, the magazine's featured articles also help the reader *browse* through the highlights. These three aspects are equally important in helping users access the content of the book or the magazine. For today's video content, techniques are urgently needed for automatically (or semiautomatically) constructing video ToC, video highlights, and video indices to facilitate summarization, browsing, and retrieval.

A great degree of power and flexibility can be achieved by simultaneously designing the video access components (ToC, highlights, and index) using a unified framework. For a long and continuous stream of data such as video, a "back and forth" mechanism between summarization and retrieval is crucial.

The rest of this chapter is organized as follows. Section 1.2 introduces important video terms. Sections 1.3 to 1.5 review video analysis, representation, and summarization and retrieval. Section 1.6 presents an overview of the rest of the book.

1.2   Terminology

Before we go into the details of the discussion, it will be beneficial to introduce some important terms used in the digital video research field:

- *Scripted/Unscripted content.* A video that is carefully produced according to a script or plan that is later edited, compiled, and distributed for consumption is referred to as scripted content. News videos, dramas, and movies are examples of scripted content. Video content that is not scripted is then referred to as unscripted content. In unscripted content, such as surveillance video, the events happen spontaneously. One can think of varying degrees of "scripted-ness" and "unscripted-ness" from movie content to surveillance content.
- *Video shot.* A consecutive sequence of frames recorded from a single camera. It is the building block of video streams.
- *Key frame.* The frame that represents the salient visual content of a shot. Depending on the complexity of the content of the shot, one or more key frames can be extracted.
- *Video scene.* A collection of semantically related and temporally adjacent shots, depicting and conveying a high-level concept or

story. While shots are marked by physical boundaries, scenes are marked by semantic boundaries.[1]

- *Video group.* An intermediate entity between the physical shots and semantic scenes and that serves as the bridge between the two. Examples of groups are temporally adjacent shots [5] or visually similar shots [3].

- *Play and break.* The first level of semantic segmentation in sports video and surveillance video. In sports video (e.g., soccer, baseball, or golf), a game is in play when the ball is in the field and the game is going on; break, or out of play, is the complement set (i.e., whenever "the ball has completely crossed the goal line or touch line, whether on the ground or in the air" or "the game has been halted by the referee") [6]. In surveillance video, a play is a period in which there is some activity in the scene.

- *Audio marker.* A contiguous sequence of audio frames representing a key audio class that is indicative of the events of interest in the video. An example of an audio marker for sports video can be the audience reaction sound (cheering and applause) or the commentator's excited speech.

- *Video marker.* A contiguous sequence of video frames containing a key video object that is indicative of the events of interest in the video. An example of a video marker for baseball videos is the video segment containing the squatting catcher at the beginning of every pitch.

- *Highlight candidate.* A video segment that is likely to be remarkable and can be identified using the video and audio markers.

- *Highlight group.* A cluster of highlight candidates.

In summary, scripted video data can be structured into a hierarchy consisting of five levels: video, scene, group, shot, and key frame, which increase in fineness of granularity from top to bottom [4] (Fig. 1.2). Similarly, the unscripted video data can be structured into a hierarchy of four levels: play/break, audio-visual markers, highlight candidates, and highlight groups, which increase in semantic level from bottom to top (Fig. 1.3).

---

[1] Some of the early literature in video parsing misused the phrase *scene change detection* for *shot boundary detection*. To avoid any later confusion, we will use *shot boundary detection* to mean the detection of physical shot boundaries and use *scene boundary detection* to mean the detection of semantic scene boundaries.

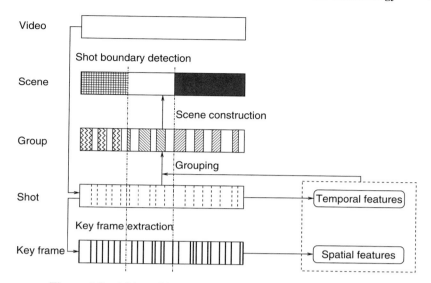

Figure 1.2   A hierarchical video representation for scripted content.

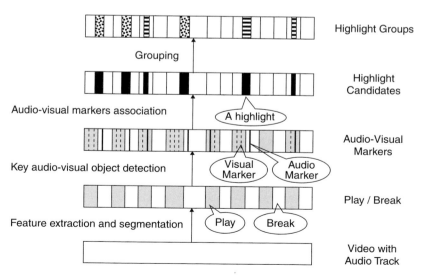

Figure 1.3   A hierarchical video representation for unscripted content.

1.3   Video Analysis

As Figure 1.1 illustrates, multimodal analysis is the basis for later video processing. It includes *shot boundary detection* and *key frame extraction* for scripted content. For unscripted content, it includes *play/break segmentation*, *audio marker detection*, and *visual marker detection*.

1.3.1   SHOT BOUNDARY DETECTION

It is not efficient (sometimes not even possible) to process a video clip as a whole. It is beneficial to first decompose the video clip into shots and do the signal processing at the shot level.

In general, automatic shot boundary detection techniques can be classified into five categories: *pixel based*, *statistics based*, *transform based*, *feature based*, and *histogram based*. Pixel-based approaches use pixel-wise intensity difference to mark shot boundaries [1,7]. However, they are highly sensitive to noise. To overcome this problem, Kasturi and Jain propose to use intensity statistics (mean and standard deviation) as shot boundary detection measures [8]. To achieve faster processing, Arman et al. propose to use the compressed discrete cosine transform (DCT) coefficients (e.g., Motion Picture Experts Group data) as the boundary measure [9]. Other transform-based shot boundary detection approaches make use of motion vectors, which are already embedded in the MPEG stream [10, 11]. Zabih et al. address the problem from another angle. Edge features are first extracted from each frame. Shot boundaries are then detected by finding sudden edge changes [12]. So far, the histogram-based approach has been the most popular. Instead of using pixel intensities directly, the histogram-based approach uses histograms of the pixel intensities as the measure. Several researchers claim that it achieves a good trade-off between accuracy and speed [1]. There are many representatives of this approach [1, 13–16]. More recent work has been based on clustering and postfiltering [17], which achieves fairly high accuracy without producing many false positives. Two comprehensive comparisons of shot boundary detection techniques are presented in Boreczky et al. and Ford et al. [18, 19].

1.3.2   KEY FRAME EXTRACTION

After the shot boundaries are detected, corresponding key frames can then be extracted. Simple approaches may just extract the first and last frames of each shot as the key frames [16]. More sophisticated key frame extraction techniques are based on visual content complexity indicators [20], shot activity indicators [21], and shot motion indicators [22, 23].

The following three analysis steps mainly cater to the analysis of unscripted content.

1.3.3   PLAY/BREAK SEGMENTATION

Since unscripted content has short periods of activity (plays) between periods of inactivity (breaks), it is useful to first segment the content into these units. This helps to reduce the amount of content to be analyzed for subsequent processing that looks for highlight segments within plays. Play/break segmentation for sports, both in an unsupervised and supervised manner using low-level features, has been reported [24]. Play/break segmentation in surveillance has been reported using adaptive background subtraction techniques that identify periods of object activity from the whole content [25].

1.3.4   AUDIO MARKER DETECTION

Audio markers are key audio classes that indicate the events of interest in unscripted content. In our previous work on sports, audience reaction and commentator's excited speech are classes that have been shown to be useful markers [26, 27]. Nepal et al. [28] detect basketball "goal" based on crowd cheers from the audio signal using energy thresholds. Another example of an audio marker, consisting of key words such as "touchdown" or "fumble," has been reported [29].

1.3.5   VIDEO MARKER DETECTION

Visual markers are key video objects that indicate the events of interest in unscripted content. Some examples of useful and detectable visual markers are "the squatting baseball catcher pose" for baseball, and "the goalpost" for soccer. Kawashima et al. [30] detect bat-swings as visual markers using visual features. Gong et al. [31] detect and track visual markers such as the soccer court, the ball, the players, and the motion patterns.

1.4   Video Representation

Considering that each video frame is a two-dimensional (2D) object and the temporal axis makes up the third dimension, a video stream spans a three-dimensional (3D) space. Video representation is the *mapping* from the 3D space to the 2D view screen. Different mapping functions characterize different video representation techniques.

1.4.1   VIDEO REPRESENTATION FOR SCRIPTED CONTENT

Using an analysis framework that can detect shots, key frames, and scenes, it is possible to come up with the following representations for scripted content.

1.4.1.1   Representation Based on Sequential Key Frames

After obtaining shots and key frames, an obvious and simple method of video representation is to sequentially lay out the key frames of the video, from top to bottom and from left to right. This simple technique works well when there are few key frames. When the video clip is long, this technique does not scale, since it does not capture the embedded information within the video clip, except for time.

1.4.1.2   Representation Based on Groups

To obtain a more meaningful video representation when the video is long, related shots are merged into groups [3, 5]. Zhang et al. [5] divide the entire video stream into multiple video segments, each of which contains an equal number of consecutive shots. Each segment is further divided into subsegments, thus constructing a tree-structured video representation. Zhong et al. [3] proposed a cluster-based video hierarchy, in which the shots are clustered based on their visual content. This method again constructs a tree-structured video representation.

1.4.1.3   Representation Based on Scenes

To provide the user with better access to the video, it is necessary to construct a video representation at the semantic level [2,4]. It is not uncommon for a modern movie to contain a few thousand shots and key frames. This is evidenced in Yeung et al. [32]—there are 300 shots in a 15-minute video segment of the movie *Terminator 2: Judgment Day*, and the movie lasts 139 minutes. Because of the large number of key frames, a simple one-dimensional (1D) sequential presentation of key frames for the underlying video (or even a tree-structured layout at the group level) is almost meaningless. More importantly, people watch the video by its semantic scenes rather than the physical shots or key frames. While *shot* is the building block of a video, it is *scene* that conveys the semantic meaning of the video to the viewers. The discontinuity of shots is overwhelmed by the continuity of a scene [2]. Video ToC construction at the scene level is thus of fundamental importance to video browsing and retrieval. In Bolle et al. [2], a scene transition graph (STG) of video representation is proposed and constructed. The video sequence is first segmented into shots. Shots are then clustered by using *time-constrained clustering*. The STG is then constructed based on the time flow of the clusters.

1.4.1.4   Representation Based on Video Mosaics

Instead of representing the video structure based on the video-scene-group-shot-frame hierarchy as discussed earlier, this approach takes a different perspective [33]. The mixed information within a shot is decomposed into three components:

- *Extended spatial information.* This captures the appearance of the entire background imaged in the shot and is represented in the form of a few mosaic images.
- *Extended temporal information.* This captures the motion of independently moving objects in the form of their trajectories.
- *Geometric information.* This captures the geometric transformations that are induced by the motion of the camera.

1.4.2   *VIDEO REPRESENTATION FOR UNSCRIPTED CONTENT*

Highlights extraction from unscripted content requires a different representation from the one that supports browsing of scripted content. This is because shot detection is known to be unreliable for unscripted content. For example, in soccer video, visual features are so similar over a long period of time that almost all the frames within it may be grouped as a single shot. However, there might be multiple semantic units within the same period such as attacks on the goal or counter attacks in the midfield. Furthermore, the representation of unscripted content should emphasize detection of remarkable events to support highlights extraction, while the representation for scripted content does not fully support the notion of an event being remarkable compared to others.

For unscripted content, using an analysis framework that can detect plays and specific audio and visual markers, it is possible to come up with the following representations.

1.4.2.1   Representation Based on Play/Break Segmentation

As mentioned earlier, play/break segmentation using low-level features gives a segmentation of the content at the lowest semantic level. By representing a key frame from each of the detected play segments, one can enable the end user to select just the play segments.

1.4.2.2   Representation Based on Audio-Visual Markers

The detection of audio and visual markers enables a representation that is at a higher semantic level than is play/break representation. Since the detected markers are indicative of the events of interest, the user can use either or both of them to browse the content based on this representation.

1.4.2.3   Representation Based on Highlight Candidates

Association of an audio marker with a video marker enables detection of highlight candidates that are at a higher semantic level. This fusion of complementary cues from audio and video helps eliminate false alarms in either of the marker detectors. Segments in the vicinity of a video marker and an associated audio marker give access to the highlight candidates for the end user. For instance, if the baseball catcher pose (visual marker) is associated with an audience reaction segment (audio marker) that follows it closely, the corresponding segment is highly likely to be remarkable or interesting.

1.4.2.4   Representation Based on Highlight Groups

Grouping of highlight candidates would give a finer resolution representation of the highlight candidates. For example, golf swings and putts share the same audio markers (audience applause and cheering) and visual markers (golfers bending to hit the ball). A representation based on highlight groups supports the task of retrieving finer events such as "golf swings only" or "golf putts only."

1.5 Video Browsing and Retrieval

These two functionalities are the ultimate goals of a video access system, and they are closely related to (and built on top of) video representations. The representation techniques for scripted content discussed earlier are suitable for browsing through ToC based summarization, while the last technique, i.e. extracting video mosaics, can be used in video retrieval. On the other hand, the representation techniques for unscripted content are suitable for browsing through highlights-based summarization.

1.5.1 VIDEO BROWSING USING TOC-BASED SUMMARY

For representation based on sequential key frames, browsing is obviously sequential browsing, scanning from the top-left key frame to the bottom-right key frame. For representation based on groups, a hierarchical browsing is supported [3,5]. At the coarse level, only the main themes are displayed. Once the user determines which theme he or she is interested in, that user can then go to the finer level of the theme. This refinement process can go on until the leaf level. For the STG representation, a major characteristic is its indication of time flow embedded within the representation. By following the time flow, the viewer can browse through the video clip.

1.5.2 VIDEO BROWSING USING HIGHLIGHTS-BASED SUMMARY

For representation based on play/break segmentation, browsing is also sequential, enabling a scan of all the play segments from the beginning of the video to the end. Representation based on audio-visual markers supports queries such as "find me video segments that contain the soccer goalpost in the left-half field," "find me video segments that have the audience applause sound," or "find me video segments that contain the squatting baseball catcher." Representation based on highlight candidates supports queries such as "find me video segments where a golfer has a good hit" or "find me video segments where there is a soccer goal attempt." Note that "a golfer has a good hit" is represented by the detection of the golfer hitting the ball followed by the detection of applause from the audience. Similarly, that "there is a soccer goal attempt" is represented by the detection of the soccer goalpost followed by the detection of long and loud audience cheering. Representation based on highlight groups supports more detailed queries than the previous representation. These queries include "find me video segments where a golfer has a good swing," "find me video segments where a golfer has a good putt," or "find me video segments where there is a good soccer corner kick."

1.5.3 VIDEO RETRIEVAL

As discussed in Section 1.1, the ToC, highlights, and index are equally important for accessing the video content. Unlike the other video representations, the mosaic representation is especially suitable for video retrieval. Three components—moving objects, backgrounds, and camera motions— are perfect candidates for a video index. After constructing such a video index, queries such as "find me a car moving like this" or "find me a conference room having that environment" can be effectively supported.

1.6   The Rest of the Book

The goals of this book are to develop novel techniques for constructing the video ToC, video highlights, and video index as well as to explain how to integrate them into a unified framework. To achieve these goals, we organize the rest of the book as follows.

Chapter 2 presents a hierarchical representation that includes key frames, shots, groups, and scenes for scripted video. Detailed algorithms are given on how to select key frames, how to cluster video frames into shots, and how to group shots into groups and scenes. Experimental results are also presented.

Chapter 3 presents a hierarchical representation that includes play/ break, audio-visual markers, highlight candidates, and highlight groups for unscripted video. Detailed algorithms are given on play/break segmentation, audio-visual markers detection, highlight candidates extraction, and highlight groups clustering. Experimental results are also presented.

While Chapter 3 discusses highlights extraction using supervised machine learning, Chapter 4 focuses on how unsupervised learning can support supervised learning for the same task. We present the state-of-the-art in unsupervised analysis for summarization and browsing. We also present a novel content-adaptive representation framework for unscripted content and show how it supports the supervised analysis in Chapter 3.

Chapter 5 presents different video indexing techniques that are based on color, texture, shape, spatial layout, and motion activity. Experimental results are also presented.

Chapter 6 presents a unique and unified framework for video summarization, browsing, and retrieval to support going back and forth between video ToC and video index for scripted video and between video highlights and video index for unscripted video. Experimental results are also presented.

Chapter 7 reviews many of the video summarization, browsing, and retrieval applications that have already used the technologies described in Chapters 2, 3, 4, 5, and 6. We also review some of the limitations of these technologies.

Chapter 8 recapitulates our coverage of video browsing and indexing techniques and discusses the state of the art. We lay out the challenges for future research and development.

Chapter 2 | Video Table-of-Content Generation

2.1 Introduction

We have seen a rapid increase of the usage of multimedia information. Of all the media types (text, image, graphic, audio, and video), video is the most challenging, as it combines all other media information into a single data stream. Owing to the decreasing cost of storage devices, higher transmission rates, and improved compression techniques, digital video is becoming available at an ever-increasing rate. Many related applications have emerged. They include video on demand, digital video libraries, distance learning, and surveillance systems [34]. The European Union recently also launched the ECHO project, which is a digital library service for historical films belonging to large national audio-visual archives (http://pc-erato2.iei.pi.cnr.it/echo).

Because of video's length and its unstructured format, efficient access to video is not an easy task. Fortunately, video is not the first long medium. Access to a book is greatly facilitated by a well-designed table of contents (ToC) that captures the semantic structure of the book. For today's video, the lack of such a ToC makes the task of browsing and retrieval inefficient, because a user searching for a particular object of interest has to use the time-consuming fast-forward and rewind functions. In fact, some video (e.g., DVD) has incorporated the concept of chapters into the video metadata. But that has to be done manually and is not practical for a large archive of digital video. Therefore, efficient automatic techniques need to be developed to construct video ToC that will facilitate user's access.

In the real world, various types of videos exist, including movies, newscasts, sitcoms, commercials, sports, and documentary videos. Some

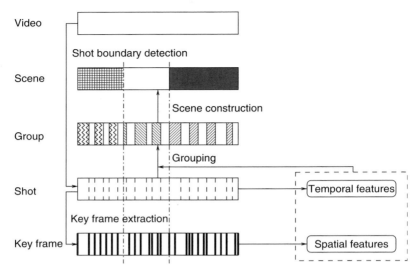

Figure 2.1   A hierarchical video representation for scripted video content.

of them, such as movies, have "story lines," while others do not (e.g., sports). In this chapter, we concentrate on videos that have story lines.

Over the past few years, researchers have implicitly used video ToC to facilitate user's access to video content, but mainly it has been limited to the shot, key frame, and group levels, for example, the representation in Figure 2.1. Since the video ToC at these levels is not closely related to the semantics of the video and normally has a large number of entries, further investigations are needed.

Take the shot and key frame–based video ToC, for example. It is not uncommon for a modern movie to contain a few thousand shots and key frames. This is evident in Yeung et al. [35]—there are 300 shots in a 15-minute video segment of the movie *Terminator 2: Judgment Day*, and the movie lasts 139 minutes. Because of the large number of key frames, a simple one-dimensional (1D) sequential presentation of key frames for the underlying video is almost meaningless. More important, people watch the video by its semantic scenes, not the physical shots or key frames. While a shot is the building block of a video, it is a scene that conveys the semantic meaning of the video to the viewers. The discontinuity of shots is overwhelmed by the continuity of a scene [2]. The video ToC construction at the scene level is thus of fundamental importance to video browsing and retrieval.

This chapter presents a novel framework for scene-level video ToC construction. It utilizes an intelligent, unsupervised clustering technique. It better models the "temporal locality" without having the "window effect," and it constructs more accurate scene structures by taking comprehensive information into account. The rest of the chapter is organized as follows. Section 2.2 reviews and evaluates related work in video ToC construction at various levels. Section 2.3 presents the proposed video ToC construction technique in detail. To put the proposed algorithm into practice, Section 2.4 utilizes techniques based on Gaussian normalization to determine the algorithm's parameters. Section 2.5 uses experiments over real-world movie video clips to validate the effectiveness of the proposed approach. Concluding remarks are given in Section 2.6.

2.2 Related Work

Work on extracting video ToC has been done at various levels (key frame, shot, group, and scene). Next, we briefly review and evaluate some of the common approaches proposed.

2.2.1 SHOT- AND KEY FRAME–BASED VIDEO TOC

In this approach, the raw video stream is first segmented into a sequence of shots by using automatic shot boundary detection techniques. Key frames are then extracted from the segmented shots. The video ToC is constructed as a sequence of the key frames. A user can access the video by browsing through the sequence of key frames. The supporting techniques of this approach, automatic shot boundary detection and key frame extraction, are summarized as follows:

- **Shot boundary detection**. In general, automatic shot boundary detection techniques can be classified into five categories: pixel based, statistics based, transform based, feature based, and histogram based. Pixel-based approaches use the pixel-wise intensity difference as the indicator for shot boundaries [1, 7]. One of its drawbacks is its sensitivity to noise. To overcome this problem, Kasturi and Jain propose the use of intensity statistics (mean and standard deviation) as the shot boundary detection measure [8]. Exploring how to achieve faster speed, Arman, Hsu, and Chiu propose the use of discrete cosine transform (DCT) coefficients in the compressed domain as the boundary measure [9]. Other transform-based shot boundary detection approaches make use of the motion vectors that are already embedded in the Motion Picture Expert Group (MPEG) stream [10, 11]. Zabih et al. address the problem from another angle. The edge features are first extracted from each frame. Shot boundaries are then detected by comparing the edge difference [12]. So far, the histogram-based approach is the most popular. Several researchers claim that it achieves good trade-off between accuracy and speed [1]. There are many representatives of this approach [1, 13–16]. Two comprehensive comparisons of various shot boundary detection techniques are in Boreczky and Rave [18] and Ford et al. [19].

- **Key frame extraction**. After the shot boundaries are detected, corresponding key frames can then be extracted. Simple approaches may just extract the first and last frames of each shot as the key frames [16]. More sophisticated key frame extraction techniques are based on the visual content complexity indicator [20], the shot activity indicator [21], and the shot motion indicator [22].

After the shot boundaries are detected and key frames extracted, the sequence of key frames, together with their frame indentifications (IDs) are used as the video ToC. This approach works well when the number of key frames is not large. However, this approach does not scale to long video clips. In those cases, a simple sequential display of key frames is almost meaningless, as discussed in Section 2.1.

2.2.2   GROUP-BASED VIDEO TOC

To obtain a video ToC at a higher level in the video representation hierarchy (Figure 2.1), related shots are merged into groups, based on which a browsing tree can be constructed [3, 5]. Zhang et al. [5] divide the whole video stream into multiple video segments, each of which contains an equal number of consecutive shots. Each segment is further divided into subsegments, thus constructing a hierarchy of video content that is used to assist browsing. In this approach, time is the only factor considered, and no visual content is used in constructing the browsing hierarchy. In contrast, Zhong et al. [3] proposed a cluster-based video hierarchy in which the shots are clustered based on their visual content. Although this approach takes into account the visual content, the time factor is lost. One of the common drawbacks of these approaches is that the video structuring is not at a high enough semantic level. Although these group-based video ToCs provide better solutions than the shot and key frame–based video ToCs, they still convey only little underlying semantic concepts and stories of the video.

2.2.3 SCENE-BASED VIDEO TOC

To provide the user with better access to the video, it is necessary to construct the video ToC at a semantic level. Existing approaches to scene-level video ToC construction can be classified into two categories: model based and general purpose. In the model-based approach, an a priori model of a particular application or domain is first constructed. This model specifies the scene boundary characteristics, based on which the unstructured video stream can be abstracted into a structured representation. The theoretical framework of this approach has been proposed by Swangberg, Shu, and Jain [14], and it has been successfully realized in many interesting applications, including news video parsing [36] and TV soccer program parsing [31]. Since this approach is based on specific application models, it normally achieves high accuracy. One of the drawbacks of this approach, however, is that for each application, a model needs to be constructed before the parsing process can proceed. The modeling process is time consuming and requires good domain knowledge and experience.

Another approach to scene-based video ToC construction does not require such an explicit domain model. Three of the pioneering works of this approach are from Institute de Recherche en Informatique de Toulouse (IRIT), France [37], Princeton University, and IBM [2, 11, 32, 35] and Toshiba Corp. [38]. Aigrain, Joly, and Longueville [37] propose a multimodal rule-based approach. They first identify local (in time) rules, which are given by the medium contents, and then construct scenes (which they call "macrosegments") by combining the rules. In [2] and [35], the video stream is first segmented into shots. Then a time-constrained clustering is used to construct visually similar and temporally adjacent shots into clusters. Finally, a scene transition graph (STG) is constructed based on the clusters, and cutting edges are identified to construct the scene structure. In Aoki et al. [38], instead of using the STG, the authors group shots of alternating patterns into scenes (which they call "acts"). A two-dimensional (2D) presentation of the video structure is then created, with scenes displayed vertically and key frames displayed horizontally.

The advantages of a scene-based video ToC over the other approaches are summarized as follows:

- The other approaches produce too many entries to be efficiently presented to the viewer.

- Shots, key frames, and even groups convey only physical discontinuities, while scenes convey semantic discontinuities, such as scene changes in time or location.

2.3   The Proposed Approach

Based on the discussion in the previous section, it is obvious that scene-based ToC has advantages over other approaches. Within the scene-based approaches, the multimodal rule-based method is not at a matured stage yet. More rules need to be generated and tested before the method can be put into practice. Furthermore, the process of generating the rules may be as time consuming as generating application models. We thus focus our attention on other approaches that have been developed [2, 35, 38].

Our proposed approach utilizes an intelligent unsupervised clustering technique for scene-level video ToC construction. It better models the "temporal locality" without having the "window effect," and it constructs more accurate scene structures by taking comprehensive information into account. It has four major modules: shot boundary detection and key frame extraction, spatiotemporal feature extraction, time-adaptive grouping, and scene structure construction. While features can include both visual and audio features [34], this chapter focuses on the visual feature. The same framework can be applied to audio and textual features.

2.3.1   SHOT BOUNDARY DETECTION AND KEY FRAME EXTRACTION

As described in Section 2.2.1, a lot of work has been done in shot boundary detection, and many of the approaches achieve satisfactory performance [39]. In this chapter, we use an approach similar to the one used in Zhang et al. [1]. For key frame extraction, although more sophisticated techniques exist [20–22], they require high computation effort. We select the beginning and ending frames of a shot as the two key frames to achieve fast processing speed.

2.3.2   SPATIOTEMPORAL FEATURE EXTRACTION

At the shot level, the shot activity measure is extracted to characterize the temporal information:

$$Act_i = \frac{1}{N_i - 1} \sum_{k=1}^{N_i-1} Diff_{k,k-1} \qquad (2.3.1)$$

$$Diff_{k,k-1} = Dist\,(Hist(k), Hist(k-1)) \qquad (2.3.2)$$

where Act_i and N_i are the activity measure and number of frames for shot i; $Diff_{k,k-1}$ is the color histogram difference between frames k and $k-1$; $Hist(k)$ and $Hist(k-1)$ are the color histograms for frames k and $k-1$; and $Dist()$ is a distance measure between histograms. We adopt the intersection distance [40] in this chapter. The color histograms used are 2D histograms along the H and S axes in HSV color space. We disregard the V component because it is less robust to the lighting condition.

At the key frame level, visual features are extracted to characterize the spatial information. In the current algorithm, color histograms of the beginning and ending frames are used as the visual features for the shot:

$$Hist(b_i) \qquad (2.3.3)$$

$$Hist(e_i) \qquad (2.3.4)$$

where b_i and e_i are the beginning and ending frames of shot i.

Based on the preceding discussion, a shot is modeled as

$$shot_i = shot_i(b_i, e_i, Act_i, Hist(b_i), Hist(e_i)) \qquad (2.3.5)$$

which captures both the *spatial* and the *temporal* information of a shot. At higher levels, this spatial-temporal information is used in grouping and scene structure construction.

2.3.3   TIME-ADAPTIVE GROUPING

Before we construct the scene structure, it is convenient to first create an intermediate entity *group* to facilitate the process. The purpose is to place similar shots into groups, since similar shots have a high possibility of being in the same scene. For shots to be similar, the following properties should be satisfied:

- **Visual similarity**. Similar shots should be visually similar. That is, they should have similar spatial ($Hist(b_i)$ and $Hist(e_i)$) and temporal (Act_i) features.
- **Time locality**. Similar shots should be close to each other temporally [35]. For example, visually similar shots, if far apart from each other in time, seldom belong to the same scene and hence not to the same group.

Yeung et al. [35] propose a *time-constrained clustering* approach to grouping shots, where the similarity between two shots is set to 0 if their time difference is greater than a predefined threshold. We propose a more general *time-adaptive grouping* approach based on the two properties for the similar shots just described. In our proposed approach, the similarity of two shots is an increasing function of visual similarity and a decreasing function of frame difference. Let i and j be the indices for the two shots whose similarity is to be determined, where shot $j >$ shot i. The calculation of the shot similarity is described as follows:

(1) Calculate the shot color similarity $ShotColorSim$:

- Calculate the four raw frame color similarities:
  $FrameColorSim_{b_j,e_i}$, $FrameColorSim_{e_j,e_i}$,
  $FrameColorSim_{b_j,e_i}$, and $FrameColorSim_{e_j,b_i}$, where
  $FrameColorSim_{x,y}$ is defined as

$$FrameColorSim_{x,y} = 1 - Diff_{x,y} \qquad (2.3.6)$$

  where x and y are two arbitrary frames, with $x > y$.

- To model the importance of time locality, we introduce the concept of *temporal attraction*, $Attr$, which is a decreasing function of the frame difference:

$$Attr_{b_j,e_i} = max\left(0, 1 - \frac{b_j - e_i}{baseLength}\right) \qquad (2.3.7)$$

$$Attr_{e_j,e_i} = max\left(0, 1 - \frac{e_j - e_i}{baseLength}\right) \qquad (2.3.8)$$

$$Attr_{b_j,b_i} = max\left(0, 1 - \frac{b_j - b_i}{baseLength}\right) \qquad (2.3.9)$$

$$Attr_{e_j,b_i} = max\left(0, 1 - \frac{e_j - b_i}{baseLength}\right) \qquad (2.3.10)$$

$$baseLength = MULTIPLE * avgShotLength \qquad (2.3.11)$$

where $avgShotLength$ is the average shot length of the whole video stream; $MULTIPLE$ is a constant that controls how fast the temporal attraction will decrease to 0. For our experiment data, we find $MULTIPLE = 10$ gives good results. The preceding definition of *temporal attraction* says that the farther apart the frames, the less the *temporal attraction*. If the frame difference is larger than $MULTIPLE$ times the average shot length, the attraction decreases to 0.

- Convert the raw similarities to *time-adaptive* similarities, which capture both the visual similarity and time locality:

$$FrameColorSim'_{b_j,e_i} = Attr_{b_j,e_i} \times FrameColorSim_{b_j,e_i}$$

$$(2.3.12)$$

$$FrameColorSim'_{e_j,e_i} = Attr_{e_j,e_i} \times FrameColorSim_{e_j,e_i}$$

$$(2.3.13)$$

$$FrameColorSim'_{b_j,b_i} = Attr_{b_j,b_i} \times FrameColorSim_{b_j,b_i}$$

$$(2.3.14)$$

$$FrameColorSim'_{e_j,b_i} = Attr_{e_j,b_i} \times FrameColorSim_{e_j,b_i}$$

$$(2.3.15)$$

- The color similarity between shots i and j is defined as the maximum of the four frame similarities:

$$ShotColorSim_{i,j} = max(FrameColorSim'_{b_j,e_i},$$
$$FrameColorSim'_{e_j,e_i},$$
$$FrameColorSim'_{b_j,b_i},$$
$$FrameColorSim'_{e_j,b_i}) \quad (2.3.16)$$

(2) Calculate the shot activity similarity:

$$ShotActSim_{i,j} = Attr_{center} \times |Act_i - Act_j| \quad (2.3.17)$$

$$Attr_{center} = max\left(0, 1 - \frac{(b_j + e_j)/2 - (b_i + e_i)/2}{baseLength}\right) \quad (2.3.18)$$

where $Attr_{center}$ is the temporal attraction between the two center frames of shot i and shot j.

(3) Calculate the overall shot similarity $Shot Sim$:

$$ShotSim_{i,j} = W_C \times ShotColorSim_{i,j} + W_A \times ShotActSim_{i,j}$$

$$(2.3.19)$$

where W_C and W_A are appropriate weights for color and activity measures.

2.3.4 SCENE STRUCTURE CONSTRUCTION

Similar shots are grouped into a group, but even nonsimilar groups can be grouped into a single scene if they are *semantically related*. Video is a sequential medium. Therefore, even though two or more processes are developing simultaneously in a video, they have to be displayed sequentially, one after another. This is common in a movie. For example, when two people are talking to each other, even though both people contribute to the conversation, the movie switches back and forth between these two people. In this example, clearly two groups exist, one corresponding to person A and the other corresponding to person B. Even though these two groups are nonsimilar, they are semantically related and constitute a single scene.

We propose an intelligent unsupervised clustering technique to perform scene structure construction. This is achieved in a two-step process:

- Collect similar shots into groups using *time-adaptive grouping*.
- Merge semantically related groups into a unified scene.

The advantages of the proposed approach over existing approaches are summarized as follows:

- *Temporal continuity*. In Yeung et al. [35], a time-window of width T is used in the *time-constrained clustering*. Similarly, in Aoki et al. [38], a search window that is eight shots long is used when calculating the shot similarities. While this "window" approach is a big advance from the plain unsupervised clustering in video analysis, it has the problem of discontinuity ("window effects"). For example, if the frame difference between two shots is $T - 1$, then the similarity between these two shots is kept unchanged. But if these two shots are a bit farther apart from each other, making the frame difference to be $T + 1$, the similarity between these shots is suddenly cleared to 0. This discontinuity may cause the clustering to be wrong and make the clustering results sensitive to the window size. To overcome this discontinuity problem, in our proposed approach, we introduce the concept of *temporal attraction*, which is a continuous and decreasing function of frame difference, as shown in Equations (2.3.7) to (2.3.10) and Equation (2.3.18). Temporal attraction effectively models the importance of time locality and does not cause any discontinuity in grouping.
- *Direct merge to a scene*. In many cases, the current shot may not be similar enough to any group in a scene; thus, it cannot be directly

merged to the scene. However, it may be similar to a certain degree to most of the groups in a scene. For example, a camera shoots three sets of shots of a person from three angles: $30°$, $60°$, and $45°$. Obviously, the three sets of shots will form three groups. Suppose the first two groups have already been formed and the current shot is a shot in group 3. While the current shot may not be similar to either of groups 1 and 2, it is similar to some extent to both groups 1 and 2; and all the three groups are semantically related. This situation occurs often in video shooting. The approaches in Yeung et al. [35] and Aoki et al. [38] will not be effective in handling this, since both of them only compare the current shot to the individual groups but not to the scene as a whole. In the proposed approach, besides calculating the similarities between the current shot and groups (Step 3 in the main procedure that follows), we also calculate the similarities between the current shot and the scene, which consists of multiple groups (Step 4 in the main procedure). This added functionality effectively takes into account the example situation and merges all three groups into a unified scene. The details of the proposed approach are described next.

[Main Procedure]

- Input: Video shot sequence, $S = \{shot\ 0, \ldots, shot\ i\}$.
- Output: Video structure in terms of *scene*, *group*, and *shot*.
- Procedure:

  (1) Initialization: assign shot 0 to group 0 and scene 0; initialize the group counter *numGroups* $= 1$; initialize the scene counter *numScenes* $= 1$.

  (2) If S is empty, quit; otherwise get the next shot. Denote this shot as shot i.

  (3) Test if shot i can be merged to an existing group:

     (a) Compute the similarities between the current shot and existing groups: Call *findGroupSim()*.

     (b) Find the maximum group similarity:

$$maxGroupSim_i = \max_g\ GroupSim_{i,g}, g = 1, \ldots, numGroups$$

$$(2.3.20)$$

where $GroupSim_{i,g}$ is the similarity between shot i and group g. Let the group of the maximum similarity be group g_{max}.

(c) Test if this shot can be merged into an existing group: If $maxGroupSim_i > groupThreshold$, where $groupThreshold$ is a predefined threshold:

   (i) Merge shot i to group g_{max}.

   (ii) Update the video structure: Call $updateGroupScene()$.

   (iii) Goto Step 2.

Otherwise:

   (i) Create a new group containing a single shot i. Let this group be group j.

   (ii) Set $numGroups = numGroups + 1$.

(4) Test if shot i can be merged to an existing scene:

   (a) Calculate the similarities between the current shot i and existing scenes: Call $findSceneSim()$.

   (b) Find the maximum scene similarity:

$$maxSceneSim_i = s_{max} \; SceneSim_{i,s}, s = 1, \ldots, numScenes$$
$$(2.3.21)$$

where $SceneSim_{i,s}$ is the similarity between shot i and scene s. Let the scene of the maximum similarity be scene s_{max}.

   (c) Test if shot i can be merged into an existing scene: If $maxSceneSim_i > sceneThreshold$, where $sceneThreshold$ is a predefined threshold:

   (i) Merge shot i to scene s_{max}.

   (ii) Update the video structure: Call $updateScene()$.

Otherwise:

   (i) Create a new scene containing a single shot i and a single group j.

   (ii) Set $numScenes = numScenes + 1$.

(5) Goto Step 2.

The input to the algorithm is an unstructured video stream, while the output is a structured video consisting of scenes, groups, shots, and key frames, based on which the video ToC is constructed (Figure 2.2).

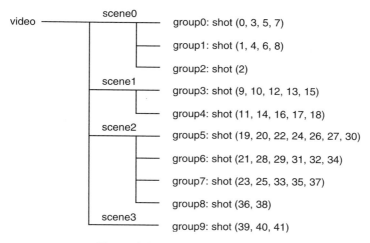

Figure 2.2   An example video ToC.

[findGroupSim]

- Input: Current shot and group structure.
- Output: Similarity between current shot and existing groups.
- Procedure:
  (1) Denote the current shot as shot i.
  (2) Calculate the similarities between shot i and existing groups:

$$GroupSim_{i,g} = ShotSim_{i,g_{last}}, g = 1, \ldots, numGroups \quad (2.3.22)$$

  where $ShotSim_{i,j}$ is the similarity between shots i and j; and g is the index for groups and g_{last} is the last (most recent) shot in group g. That is, the similarity between the current shot and a group is the similarity between the current shot and the most recent shot in the group. The most recent shot is chosen to represent the whole group because all the shots in the same group are visually similar and the most recent shot has the largest *temporal attraction* to the current shot.
  (3) Return.

[findSceneSim]

- Input: The current shot, group structure, and scene structure.
- Output: Similarity between the current shot and existing scenes.

- Procedure:
  (1) Denote the current shot as shot i.
  (2) Calculate the similarity between shot i and existing scenes:

$$SceneSim_{i,s} = \frac{1}{numGroups_s} \sum_{g}^{numGroups_s} GroupSim_{i,g} \quad (2.3.23)$$

where s is the index for scenes; $numGroups_s$ is the number of groups in scene s; and $GroupSim_{i,g}$ is the similarity between current shot i and g^{th} group in scene s. That is, the similarity between the current shot and a scene is the average of similarities between the current shot and all the groups in the scene.
  (3) Return.

[updateGroupScene]

- Input: Current shot, group structure, and scene structure.
- Output: An updated version of group structure and scene structure.
- Procedure:
  (1) Denote the current shot as shot i and the group having the largest similarity to shot i as group g_{max}. That is, shot i belongs to group g_{max}.
  (2) Define two shots, *top* and *bottom*, where *top* is the second most recent shot in group g_{max} and *bottom* is the most recent shot in group g_{max} (i.e., the current shot).
  (3) For any group g, if any of its shots (*shot g_j*) satisfies the following condition

$$top < shot\ g_j < bottom \quad (2.3.24)$$

merge the scene that group g belongs to into the scene that group g_{max} belongs to. That is, if a scene contains a shot that is interlaced with the current scene, merge the two scenes. This is illustrated in Figure 2.3 (shot $i = $ shot 4, $g_{max} = 0$, $g = 1$, *top =* shot 1, and *bottom =* shot 4).
  (4) Return.

[updateScene]

- Input: Current shot, group structure, and scene structure.
- Output: An updated version of scene structure.

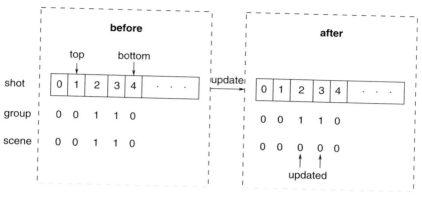

Figure 2.3   Merging scene 1 to scene 0.

- Procedure:
    (1) Denote the current shot as shot i and the scene having the largest similarity to shot i as scene s_{max}. That is, shot i belongs to scene s_{max}.
    (2) Define two shots, *top* and *bottom*, where *top* is the second most recent shot in scene s_{max} and *bottom* is the current shot in scene s_{max} (i.e., current shot).
    (3) For any scene s, if any of its shots (*shot s_j*) satisfies the following condition

    $$top < shot\ s_j < bottom \qquad (2.3.25)$$

    merge scene s into scene s_{max}. That is, if a scene contains a shot that is interlaced with the current scene, merge the two scenes.
    (4) Return.

What distinguishes the proposed approach from the plain clustering-based approach is the intelligence involved. The intelligence manifests itself in the algorithm in three aspects. First, the "temporal attraction" is an intelligent way of modeling the temporal factor of the similarity. Second, *updateGroupScene* intelligently merges related groups to a single scene. Finally, *updateScene* intelligently updates related scenes into a unified one.

The procedure *updateGroupScene* and *updateScene* are of significant importance to the proposed scene structure construction algorithm. While *findGroupSim* helps to group similar shots into a group and *findSceneSim* helps to merge a shot (or a single-element group) into a scene, it is *updateGroupScene* and *updateScene* that link semantically related shots into a single scene. For example, for scene 0 in Figure 2.2, while *findGroupSim* helps to group shots 0, 3, 5, 7 into group 0, and *findSceneSim* helps to group shot 2 to scene 0, it is *updateGroupScene* and *updateScene* that link all three groups into one unified scene.

2.4 Determination of the Parameters

There are four parameters in the proposed video ToC construction algorithm: W_C, W_A, *groupThreshold*, and *sceneThreshold*. For any algorithm to be of practical use, all the parameters should be determined either automatically by the algorithm itself or easily by the user. In our proposed algorithm, Gaussian normalization is used in determining the four parameters. Specifically, W_C, and W_A are determined automatically by the algorithm, and *groupThreshold* and *sceneThreshold* are determined by user's interaction.

2.4.1 GAUSSIAN NORMALIZATION

In Equation (2.3.19), we combine color histogram similarity and activity similarity to form the overall shot similarity. Since the color histogram feature and activity feature are from two totally different physical domains, it would be meaningless to combine them without normalizing them first. The Gaussian normalization process ensures that entities from different domains are normalized to the same dynamic range. The normalization procedure is described as follows:

[findMeanAndStddev]

- Input: Video shot sequence, $S = \{shot_0, \cdots, shot_i\}$, and a feature F associated with the shots. For example, the feature F can be either a color histogram feature or an activity feature.
- Output: The mean μ and standard deviation σ of this feature F for this video.
- Procedure:
  (1) If S is not empty, get the next shot; otherwise, goto Step 3.
  (2) Denote the current shot as shot i.
      (a) Compute the similarity in terms of F between shot i and shot i', $i' = i - MULTIPLE, \cdots, i - 1$. Note that only the similarities of the previous *MULTIPLE* shots need to be calculated, since shots outside *MULTIPLE* have zero *temporal attraction* to the current shot.

(b) Store the calculated similarity values in an array A_s.

(c) Goto Step 1.

(3) Let N_A be the number of entries in array A_s. Consider this array as a sequence of Gaussian variables, and compute the mean μ_{A_s} and standard deviation σ_{A_s} of the sequence.

The means and standard deviations for color histograms and activity measures are first calculated (denoted as μ_C, σ_C, μ_A, and σ_A) by the previous normalization procedure before the scene construction procedure is applied in Section 2.3.4. During the scene construction procedure, μ_C, σ_C, μ_A, and σ_A are used to convert the raw similarity values to normalized ones. That is, Step 3 in Section 2.3.4 (Equation (2.3.19)) is modified as follows:

(4) Calculate the overall shot similarity:

$$ShotSim_{i,j} = W_C \times ShotColorSim'_{i,j} + W_A \times ShotActSim'_{i,j}$$

$$(2.4.1)$$

$$ShotColorSim'_{i,j} = \frac{ShotColorSim_{i,j} - \mu_C}{\sigma_C}$$

$$(2.4.2)$$

$$ShotActSim'_{i,j} = \frac{ShotActSim_{i,j} - \mu_A}{\sigma_A}$$

$$(2.4.3)$$

where W_C and W_A are appropriate weights for color and activity measures; and $ShotColorSim_{i,j}$ and $ShotActSim_{i,j}$ are the raw similarity values. This procedure converts the raw similarities into similarities that obey the normal distribution of $N(0, 1)$. Being of the same distribution, the normalized color histogram similarity and the normalized activity similarity can be meaningfully combined into an overall similarity. How to determine the appropriate values for W_C and W_A is discussed in the following subsection.

2.4.2 DETERMINING W_C AND W_A

After the Gaussian normalization procedure, the raw similarities of both color histogram and activity measure are brought into the same dynamic range. That is, the normalized similarities of the color histogram feature and the activity feature contribute equally to the overall similarity. To reflect the relative importance of each feature, different weights are then associated with the features.

The relative importance of a feature can be estimated from the statistics of its feature array A_s. For example, if all the elements in A_s are of similar value, then this particular feature is of little discriminating power and should receive low weight. On the other hand, if the elements in A_s demonstrate variation, then the feature has good discriminating power and should receive high weight. Based on this intuition, the standard deviation of the feature array A_s furnishes a good estimation of the feature's importance (weight). In our case, W_C and W_A can be automatically determined as follows:

$$W_C = \frac{\sigma_C}{\sigma_C + \sigma_A} \tag{2.4.4}$$

$$W_A = \frac{\sigma_A}{\sigma_C + \sigma_A} \tag{2.4.5}$$

where σ_C and σ_A are obtained from the procedure findMeanAndStddev.

2.4.3   DETERMINING GROUPTHRESHOLD AND SCENETHRESHOLD

The *groupThreshold* and *sceneThreshold* are two important parameters in the proposed algorithm. The determination of these two thresholds would be difficult and time consuming if the shot similarities were not normalized, since in that case the thresholds are as follows:

- **Feature dependent.** Different features (color histogram versus activity measure) may require different thresholds.
- **Case dependent.** Different videos may require different thresholds.

But after the Gaussian normalization procedure, the similarity distribution of any feature for any video is normalized to the Gaussian $N(0, 1)$ distribution, making the determination of thresholds much easier. The Gaussian $N(0, 1)$ distribution is plotted in Figure 2.4.

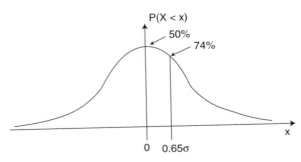

Figure 2.4   The Gaussian $N(0, 1)$ distribution.

A learning process can be designed to find appropriate values for the two thresholds. Since any feature's similarity in any video is mapped to the same distribution, the training feature and video can be arbitrary. During the training, humans manually adjust the two thresholds to obtain good video scene structure. This learning process needs to be done only once. The determined two thresholds can then be used for any other features in any other videos. Experimentally we find that $groupThreshold = 0.65 \times \sigma = 0.65$ and $sceneThreshold = 0$ gives good scene structure. This set of thresholds is used throughout of the experiments reported in Section 2.5. Note that $sceneThreshold = 0$ gives good scene structure. This set of $P(X < x|x = groupThreshold = 0.65) = 0.74$ and $P(X < x|x = sceneThreshold = 0) = 0.5$. These two probabilities indicate the following:

- Only if a shot is very similar to a group (better than 74% of all the shots) can it be merged to the group.
- When a shot is similar to some extent (better than 50% of all the shots) to all the groups in a scene, it is considered similar to the scene, which matches the physical meaning of the two thresholds, as discussed in Section 2.3.4.

2.5   Experimental Results

In all of the experiments reported in this section, the video streams are MPEG compressed, with the digitization rate equal to 30 frames per second. To validate the effectiveness of the proposed approach, representatives of various movie types are tested. Specifically, the test set includes Movie1 (romantic-slow), Movie2 (romantic-fast), Movie3 (music), Movie4 (comedy), Movie5 (science fiction-slow), Movie6 (science fiction-fast), and Movie7 (action). Each video clip is about 10 to 20 minutes long, and the total length is about 175,000 frames. The experimental results are shown in Table 2.1, where "ds" denotes the number of scenes detected by the algorithm; "fn" indicates the number of scenes missed by the algorithm; and "fp" indicates the number of scenes detected by the algorithm but not considered as scenes by humans.

Since scene is a semantic-level concept, the ground truth of scene boundary is not always concrete, and this might be the reason that some authors [2, 35, 38] do not include the two columns of "false negative" and "false positive" in their experimental result tables. But to judge the

Table 2.1   **Scene structure construction results.**

Movie Name	Frames	Shots	Groups	ds	fn	fp
Movie1	21717	133	16	5	0	0
Movie2	27951	186	25	7	0	1
Movie3	14293	86	12	6	1	1
Movie4	35817	195	28	10	1	2
Movie5	18362	77	10	6	0	0
Movie6	23260	390	79	24	1	10
Movie7	35154	329	46	14	1	2

effectiveness of a video ToC construction approach, we believe it is useful to include those two columns. Although scene is a semantic concept, relative agreement can be reached among different people. The ground truth of the scene boundaries for the tested video sequences are obtained from subjective tests. Multiple human subjects are invited to watch the movies and then are asked to give their own scene structures. The structure that most people agree with is used as the ground truth of the experiments.

Some observations can be made from the results in Table 2.1:

- The proposed scene construction approach achieves reasonably good results in most of the movie types.
- The approach achieves better performance in the "slow" movies than in the "fast" movies. This is because in the "fast" movies, the visual content is normally more complex and more difficult to capture. We are currently integrating closed-caption information into the framework to enhance the accuracy of the scene structure construction.
- The proposed approach seldom misses a scene boundary but tends to oversegment the video. That is, the number of the "false positive" is

more than that of the "false negative." This situation is expected for most of the automated video analysis approaches and has also been observed by other researchers [2, 35].

- The proposed approach is practical in terms of parameter determination. A single set of thresholds ($groupThreshold =$ $0.65 \times \sigma = 0.65$ and $sceneThreshold = 0$) is used throughout the experiments, and W_C and W_A are determined automatically by the algorithm itself.

A possible remedy to the drawbacks in the stated segmentation scheme can be eliminated by introducing the audio component of video into the scene change scheme [99, 100]. This introduces the concept of a documentary scene, which is inferred from the National Institute of Standards and Technology (NIST)–produced documentary videos. It uses an audio-visual score value (which is a weighted combination of an audio score and a video score) to divide the video into a set of documentary scenes. The audio score and the visual score are generated by procedures evolved out of the observations that we make on the video data.

This scheme is illustrated in Figure 2.5. Given a video, the proposed approach first generates a visual pattern and an audio pattern, respectively, based on similarity measures. It also makes use of the similarity in image background of the video frames for scene analysis [101]. The information collected from visual-, audio-, and background-based scene analysis will be integrated for audio-visual fusion. The scene change detector is composed of two main steps: adaptive scheme and redundancy check. In the adaptive scheme, the audio-visual score within an interval is first evaluated. This interval will be adaptively expanded or shrunk until a local maximum is found. In redundancy check, we eliminate the redundant scenes by a merging process.

The scene structure information (Fig. 2.2), together with the representative frames, leads to a semantic-level ToC to facilitate user's access to the video. The scene-level ToC is illustrated in Figure 2.6. Five scenes are created from the 21717-frame video clip (Movie1). By looking at the representative frames and the text annotation, the user can easily grasp the global content of the video clip. He or she can also expand the video ToC into more detailed levels, such as groups and shots (see Figure 2.7).

After browsing the hierarchical video ToC (scenes, groups, shots, etc.), the user can quickly position the portion of the video that he or she is interested in. Then, by clicking the Display button, the user can view that

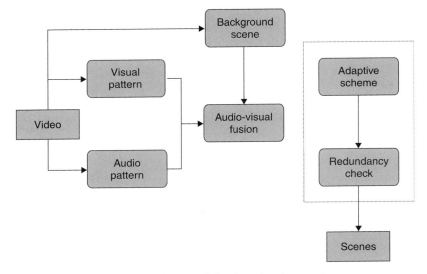

Figure 2.5   Proposed approach for detecting documentary scenes.

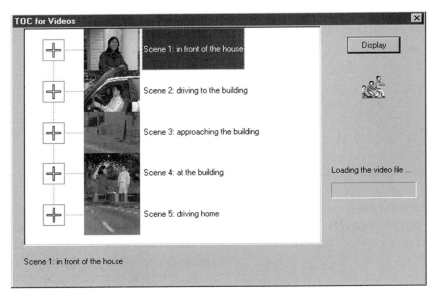

Figure 2.6   Video ToC for Movie1 (scene level).

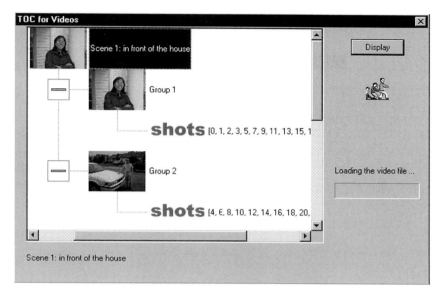

Figure 2.7   Video ToC for Movie1 (group level).

particular portion of video, without resorting to the tedious fast-forward and rewind functions.

Imagine the situation if we do not have a ToC for a long book. It may take hours to grasp the main content. The same is true for videos. The scene-based video ToC just described greatly facilitates the user's access to the video. It not only provides the user with nonlinear access to the video (in contrast to conventional linear fast-forward and rewind), but it also gives the user a global picture of the whole story.

If, instead, we were using a 1D array of key frames to present the same video, $2 \times 133 = 266$ frames have to be presented sequentially. Because of the 1D linear display nature, even if a user can patiently browse through all the 266 frames, it is still difficult for him or her to perceive the underlying story structure.

2.6   Conclusions

Motivated by the important role of the table of contents in a book, in this chapter we introduced the ToC into the video domain. We reviewed and evaluated the video parsing techniques at various levels (shots, groups, scenes). We concluded that the scene-level ToC has the advantage over other techniques. We then presented an effective scene-level ToC construction technique based on intelligent, unsupervised clustering. It has the characteristics of better modeling the time locality and scene structure. Experiments over real-world movie videos validate the effectiveness of the proposed approach. Examples are given to demonstrate the use of the scene-based ToC to facilitate the user's access to the video. The proposed approach provides an open framework for structure analysis of video—features other than the ones used in this chapter can be readily incorporated for the purpose of video ToC construction (e.g., [34]). We envision that an appropriate fusion of multiple features/modalities can result in a more semantically correct video ToC.

Chapter 3 | Highlights Extraction from Unscripted Video

3.1 Introduction

In this chapter, we describe our proposed approach for highlights extraction from "unscripted" content such as sports video. We show the effectiveness of the framework in three different sports: soccer, baseball, and golf. Our proposed framework is summarized in Figure 3.1. There are four major components in Figure 3.1. We describe them one by one in the following subsections.

3.1.1 AUDIO MARKER RECOGNITION

Broadcast sports content usually includes audience reaction to the interesting moments of the games. Audience reaction classes, including applause, cheering, and the commentator's excited speech, can serve as audio markers. We have developed classification schemes that can achieve very high recognition accuracy on these key audio classes [26]. Figure 3.2 shows our proposed unified audio marker detection framework for sports highlights extraction. More details can be found in Section 3.2.

3.1.2 VISUAL MARKER DETECTION

As defined earlier, visual markers are key visual objects that indicate the interesting segments. Figure 3.3 shows examples of some visual markers for three different games. For baseball games, we want to detect the pattern in which the catcher squats waiting for the pitcher to pitch the ball; for golf games, we want to detect the players bending to hit the golf ball; and for

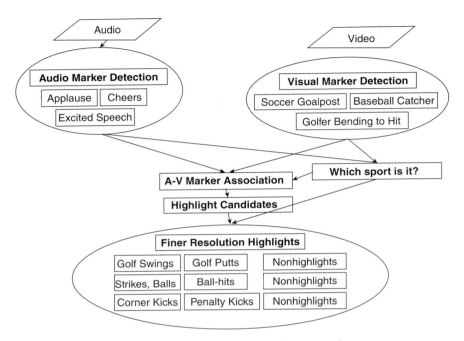

Figure 3.1   Our proposed framework: an overview.

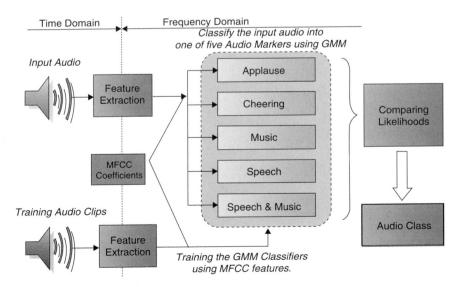

Figure 3.2   Audio markers for sports highlights extraction.

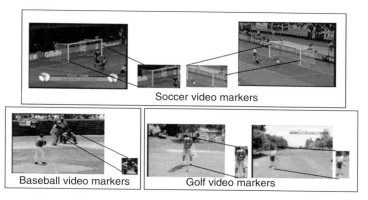

Figure 3.3 Examples of visual markers for different sports.

soccer, we want to detect the appearance of the goalpost. Correct detection of these key visual objects can eliminate the majority of the video content that is not in the vicinity of the interesting segments. *For the goal of one general framework for all three sports, we use the following processing strategy: for the unknown sports content, we detect whether there are baseball catchers or golfers bending to hit the ball, or soccer goalposts. The detection results can enable us to decide which sport (baseball, golf, or soccer) is being played.* More details can be found in Section 3.3.

3.1.3 AUDIO-VISUAL MARKER ASSOCIATION AND FINER-RESOLUTION HIGHLIGHTS

Ideally each visual marker can be associated with one and only one audio marker and vice versa. Thus, they make a pair of audio-visual markers indicating the occurrence of a highlight event in their vicinity. But since many pairs might be wrongly grouped due to false detections and misses, some postprocessing is needed to keep the error to a minimum.

Highlight candidates, delimited by the audio markers and visual markers, are quite diverse. For example, golf swings and putts share the same audio markers (audience applause and cheering) and visual markers (golfers bending to hit the ball). Both kinds of golf highlight events can be found by the aforementioned audio-visual marker detection method. To support the task of retrieving finer events such as golf swings only or golf putts only, we have developed techniques that model these events using low-level audio-visual features. Furthermore, some of these candidates might not be true highlights. We eliminate these false candidates using a finer-level highlight classification method. For example, for golf, we build models for golf swings, golf putts, and nonhighlights (neither swings nor putts) and use these models for highlights classification (swings or putts) and verification (highlights or nonhighlights). More details can be found in Section 3.4.

3.2   Audio Marker Recognition

We present an audio marker recognition approach that is based on audio classification using Gaussian mixture models (GMMs). Since the number of Gaussian components is not known a priori for any of the classes, traditionally this number is assumed to be the same for all the GMMs. This number is usually chosen through cross-validation. The practical problem is that for some class, this number will lead to overfitting of the training data if it is much less than the actual one, or, inversely, underfitting of the data. Our solution is to use the minimum description length (MDL) criterion in selecting the number of mixtures. MDL-GMMs fit the training data to the generative process as closely as possible, avoiding the problem of overfitting or underfitting.

3.2.1   *ESTIMATING THE NUMBER OF MIXTURES IN GMMs*

3.2.1.1   **Theoretical Derivations**

The derivations here follow those in Bouman [41]. Let Y be an M-dimensional random vector to be modeled using a Gaussian mixture distribution. Let K denote the number of Gaussian mixtures, and we use the notations π, μ, and R to denote the parameter sets $\{\pi_k\}_{k=1}^{K}$, $\{\mu_k\}_{k=1}^{K}$, and $\{R_k\}_{k=1}^{K}$, respectively, for mixture coefficients, means, and variances. The complete set of parameters are then given by K and $\theta = (\pi, \mu, R)$. The log of the probability of the entire sequence $Y = \{Y_n\}_{n=1}^{N}$ is then given by

$$\log p_y(y|K, \theta) = \sum_{n=1}^{N} \log \left(\sum_{k=1}^{K} p_{y_n|x_n}(y_n|k, \theta)\pi_k \right) \qquad (3.2.1)$$

The objective is then to estimate the parameters K and $\theta \in \Omega^{(K)}$. The maximum likelihood (ML) estimate is given by

$$\hat{\theta}_{ML} = \arg\max_{\theta \in \Omega^{(K)}} \log p_y(y|K, \theta)$$

the estimate of K is based on the minimization of the expression

$$MDL(K, \theta) = -\log p_y(y|K, \theta) + \frac{1}{2}L \log(NM) \qquad (3.2.2)$$

where L is the number of continuously valued real numbers required to specify the parameter θ. In this application,

$$L = K\left(1 + M + \frac{(M+1)M}{2}\right) - 1$$

Notice that this criterion has a penalty term on the total number of data values NM, what Rissanen [42] called the MDL estimator. Let us denote the parameter learning of GMMs using the MDL criterion MDL-GMM.

While the expectation maximization (EM) algorithm can be used to update the parameter θ, it does not solve the problem of how to change the model order K. Our approach will be to start with a large number of clusters and then sequentially decrement the value of K. For each value of K, we will apply the EM update until we converge to a local minimum of the MDL functional. After we have done this for each value of K, we may simply select the value of K and corresponding parameters that resulted in the smallest value of the MDL criterion.

The question remains of how to decrement the number of clusters from K to $K-1$. We will do this by merging the two closest clusters to form a single cluster. More specifically, the two clusters l and m are specified as a single cluster (l, m) with prior probability, mean, and covariance given by

$$\pi^*_{(l,m)} = \bar{\pi}_l + \bar{\pi}_m \qquad (3.2.3)$$

$$\mu^*_{(l,m)} = \frac{\bar{\pi}_l \bar{\mu}_l + \bar{\pi}_m \bar{\mu}_m}{\bar{\pi}_l + \bar{\pi}_m} \qquad (3.2.4)$$

$$R^*_{(l,m)} = \frac{\bar{\pi}_l\left(\bar{R}_l + (\bar{\mu}_l - \mu_{(l,m)})(\bar{\mu}_l - \mu_{(l,m)})^t\right)}{\bar{\pi}_l + \bar{\pi}_m}$$
$$+ \frac{\bar{\pi}_m\left(\bar{R}_m + (\bar{\mu}_m - \mu_{(l,m)})(\bar{\mu}_m - \mu_{(l,m)})^t\right)}{\bar{\pi}_l + \bar{\pi}_m} \qquad (3.2.5)$$

Here the $\bar{\pi}$, $\bar{\mu}$, and $\bar{R}$ are given by the EM update of the two individual mixtures before they are merged.

3.2.1.2   An Example: MDL-GMM for Different Sound Classes

We have collected 679 audio clips from TV broadcasts of golf, baseball, and soccer games. This database is a subset of that in Xiong et al. [43]. Each clip is hand-labeled into one of the five classes as ground truth: applause, cheering, music, speech, and speech with music. The corresponding numbers of clips are 105, 82, 185, 168, and 139. The duration of the clips differs from around 1 s to more than 10 s. The total duration is approximately 1 h and 12 min. The audio signals are all monochannel with a sampling rate of 16 kHz.

We extract 100 12-dimensional mel-frequency cepstrum coefficients (MFCC) per second using a 25-ms window. We also add the first- and second-order time derivatives to the basic MFCC parameters in order to enhance performance. For more details on MFCC feature extraction, please see Young et al. [44].

For each class of sound data, we first assign a relatively large number of mixtures to K, calculate the MDL score $MDL(K, \theta)$ using all the training sound files, then merge the two nearest Gaussian components to get the next MDL score $MDL(K - 1, \theta)$, then iterate till $K = 1$. The "optimal" number K is chosen as the one that gives the minimum of the MDL scores. For the training database we have, the relationship between $MDL(K, \theta)$ and K for all five classes are shown in Figure 3.4.

From Figure 3.4, we observe that the optimal mixture numbers of these five audio classes are 2, 2, 4, 18, and 8, respectively. This observation can be intuitively interpreted as follows. Applause or cheering has a relatively simpler spectral structure, hence fewer Gaussian components can model the data well. In comparison, because speech has a much more complex, variant spectral distribution, it needs many more components. Also, we observe that the complexity of music is between that of applause or cheering and speech. For speech with music (a mixture class of speech and music), its complexity is between the two classes that are in the mixture.

3.2.2   EVALUATION USING THE PRECISION-RECALL CURVE

Using the length L of contiguous applause or cheering as a parameter, we can count the number of true highlights N in the C candidates that have applause or cheering length greater than or equal to L. Precision is defined as the ratio between N and C (i.e., $Pr = \frac{N}{C}$). On the other hand, if we denote the total number of highlights in the ground truth set

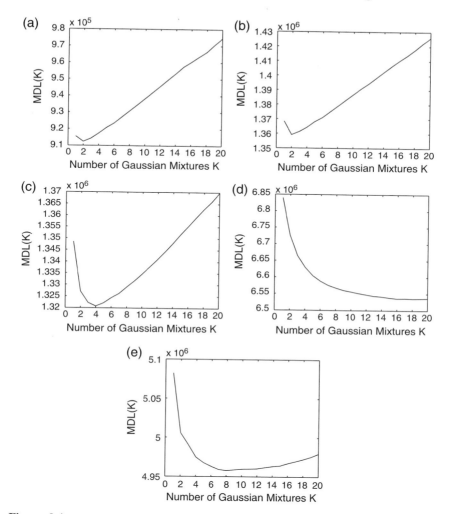

Figure 3.4 $MDL(K, \theta)$(Y-axis) with respect to different numbers of GMM mixtures K(X-axis) to model (a) applause, (b) cheering, (c) music, (d) speech, and (e) speech with music sound shown in the raster-scan order. $K = 1 \cdots 20$. The optimal mixture numbers at the lowest positions of the curves are 2, 2, 4, 18, and 8, respectively.

as G, then recall is defined as the ratio between N and G (i.e., $Re = \frac{N}{G}$). Plot Pr against Re for all possible value of L and we get the precision-recall (PR) curve that is frequently used in multimedia retrieval research. In general, a curve that is closer to the upper-right corner suggests better performance.

3.2.3 *PERFORMANCE COMPARISON*

To compare GMMs with MDL-GMMs, we cross-validate the classification results by dividing the above-mentioned five-class audio data set into 90%/10% training/test sets. For one, the number of Gaussian mixtures is assumed to be 10 for all the classes, and the test results are put into Table 3.1. For the other, the number of mixtures is composed of those chosen from Figure 3.4, and the results are shown in Table 3.2. Note that the overall classification accuracy has been improved by more than 8%.

Table 3.1 **Performance of traditional GMM, every class is modeled using 10 Gaussian mixtures: (1) applause, (2) cheering, (3) music, (4) speech, and (5) "speech with music." Classification accuracy on the 10% data by models trained on the 90% data.**

	(1)	(2)	(3)	(4)	(5)
(1)	88.8%	5.0%	3%	2%	1.2%
(2)	5%	90.1%	2%	0	2.9%
(3)	5.6%	0	88.9%	5.6%	0
(4)	0	0	0	94.1%	5.9%
(5)	0	0	6.9%	5.1%	88%

Average Recognition Rate: 90%

Table 3.2 **Performance of MDL-GMM. Classification accuracy on the 10% data by models trained on the 90% data. (1) to (5) are the same as described in Table 3.1.**

	(1)	(2)	(3)	(4)	(5)
(1)	97.1%	0	0	0.9%	2.0%
(2)	0	99.0%	1.0%	0	0
(3)	0	1.0%	99.0%	0	0
(4)	0	0	0	99.0%	1.0%
(5)	0	0	1.0%	0	99.0%

Average Recognition Rate: 98.6%

3.2.4   EXPERIMENTAL RESULTS ON GOLF HIGHLIGHTS GENERATION

We have reported some results of sports highlights extraction based on audio classification and the correlation between the applause/cheering sound with exciting moments [43]. However, there we have not used the MDL criterion to select the model structures, so we have not used the "optimal" models. Now equipped with the MDL-GMMs and with the observation that they can greatly improve classification accuracy, we revisit the problem in Xiong et al. [43].

First, instead of training on 90% and testing on 10% of the data as in Table 3.2, we train the MDL-GMMs on all the data in the ground truth set. To gain a better understanding of classification, especially on the applause/cheering sound, we also test all the data in the ground truth set before we test the game data. The results are organized into Tables 3.3 and 3.4.

Table 3.3   **The confusion matrix on *all* the audio data. The results are based on MDL-GMMs with different "optimal" numbers of mixtures (see Fig. 3.4).**

	(1)	(2)	(3)	(4)	(5)
(1)	97.1%	0	0	0.9%	2.0%
(2)	0	99.0%	1.0%	0	0
(3)	1.0%	8.0%	89.0%	0	2.0%
(4)	0	0	0	92.2%	7.8%
(5)	0	0	0.7%	2.8%	96.5%

Average Recognition Rate: 94.76%

Table 3.4   **Recognition results of those in Table 3.3. The number of sound examples that are correctly or incorrectly classified for each class is shown.**

	(1)	(2)	(3)	(4)	(5)
(1)	102	0	0	1	2
(2)	0	81	1	0	0
(3)	2	15	164	0	4
(4)	0	0	0	155	13
(5)	0	0	1	4	134

Table 3.5 Number of contiguous applause segments and highlights found by the MDL-GMMs in the golf game. These highlights are in the vicinity of the applause segments. These numbers are plotted in the left on Figure 3.5.

Applause Length	Number of Instances	Number of Highlights
$L \geq 9s$	1	1
$L \geq 8s$	3	3
$L \geq 7s$	6	5
$L \geq 6s$	11	9
$L \geq 5s$	19	15
$L \geq 4s$	35	28
$L \geq 3s$	61	43
$L \geq 2s$	101	70
$L \geq 1s$	255	95

The classification accuracy on either applause or cheering has been quite high.

We ran audio classification on the audio sound track of a 3-hour golf game (British Open, 2002). The game took place on a rainy day, so the existence of the sound of rain has corrupted our previous classification results [43] to a great degree. Every second of the game audio is classified into one of the five classes. Those contiguous applause segments are sorted according to the duration of contiguity. The distribution of these contiguous applause segments is shown in Table 3.5. Note that the applause segments can be as long as 9 continuous seconds.

Based on when the beginning of applause or cheering is, we choose to include a certain number of seconds of video before the beginning moment to include the play action (golf swings, putts, etc.), then we compare these segments to those ground truth highlights that are labeled by human viewers.

3.2.4.1 Performance and Comparison in Terms of Precision-Recall

We analyze the extracted highlights that are based on those segments in Table 3.5. For each length L of the contiguous applause segments, we calculate the precision and recall values. (Precision is the percentage of

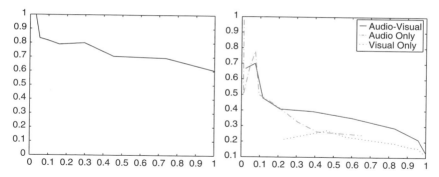

Figure 3.5 Precision-recall curves for the test golf game. Left: by the current approach; right: by the previous approaches; Y-axis: precision; X-axis: recall.

highlights that are correct of all those extracted. Recall is the percentage of highlights that are in the ground truth set.) We then plot the precision versus recall values for all different values of L into the left of Figure 3.5.

In comparison, the right-hand side of Figure 3.5 shows the results reported on the same game [45], where the dashed-line curve shows the precision-recall relationship when a hidden Markov model (HMM) is used to model the highlights using the audio class labels as the observations. These labels are generated by the models in Table 3.1. The intention of using the HMM on top of the GMM is to enhance performance. However, this is carried out for every chunk of audio data (i.e., a 12-s long window moving 1 s at a time). The solid-line curve shows the results when a coupled HMM is used to model both audio and video classes to further enhance performance on the dashed-line curve. Although we have established the argument on the superiority of coupled HMM over audio-only HMM or video-only HMM (the dotted curve), overall the performances there are not satisfactory as the best one (coupled HMM) has poor performance at the rightmost part of the curve. The reason is that we have not taken advantage of the fact that key audio classes such as applause or cheering are more indicative of the possible highlights events than other audio classes.

From the two figures in Figure 3.5, we observe that the MDL-GMMs outperform those approaches in Xiong et al. [45] by a large margin. For example, at 90% recall, the left-hand figure shows $\geq 70\%$ precision rate, while the right-hand figure shows only $\sim 30\%$ precision rate, suggesting that the false alarm rate is much lower using the current approach.

3.2.4.2 System Interface

One important application of highlight generation from sports videos is to provide viewers with the correct entry points to the video content so that they can adaptively choose other interesting content that is not necessarily modeled by the training data. This requires a progressive highlight-generation process. Depending on how long the sequence of highlights the viewers want to watch, the system should provide those most likely sequences. We thus use a content-adaptive threshold, the lowest of which is the smallest likelihood and the highest of which is the largest threshold over all the test sequences.

Then given such a time budget, we can calculate the value of the threshold above which the total length of highlight segments will be as close to the budget as possible. Then we can play those segments with likelihood greater than the threshold one after another until the budget is exhausted.

Our system interface is shown in Figure 3.6. The spiky curve at the lower half of the figure is a plot of confidence level with respect to time (second by second). Larger confidence-level values indicate more likely that there

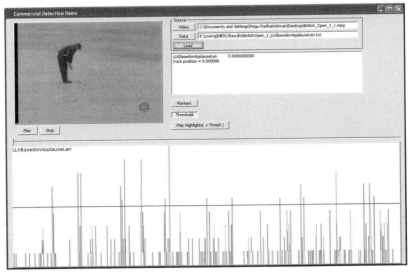

Figure 3.6   The interface of our system displaying sports highlights. The horizontal line imposed on the curve is the threshold value the user can choose to display those segments with a confidence level greater than the threshold.

are highlights at the time instance. We provide a moving threshold for the user to place on the curve. Those segments with values greater than the threshold are played one after another.

In comparison, the curve in Xiong et al. [45], which is reproduced in Figure 3.7, is much less spiky. This suggests that the approach based on audio marker recognition as presented in this chapter outperforms the one that is based on "A/V features + CHMMs" [45].

We have demonstrated the importance of a better understanding of model structures in the audio analysis for sports highlights generation. In fact, we looked into model parameter selection in terms of number of states of the HMMs: coupled HMMs (CHMMs) [45]. We showed there that the selection also improved recognition accuracy. What we have presented in this chapter is complementary to that in the following sense: MDL-GMMs can find better GMM structures; the techniques in Xiong et al. [45] can find better HMM structures.

One key factor for this poor performance is that since the boundaries of the highlights segments embedded in the entire broadcast sports content are unknown, we have used an exhaustive search strategy—that is, testing on every moving video chunk (12-s long hopping 1 s at a time). This

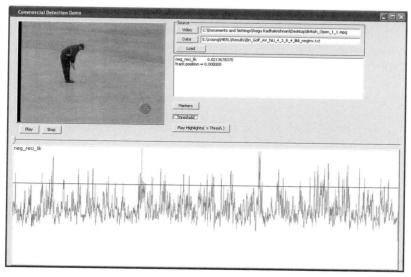

Figure 3.7 A snapshot of the interface of the results in our previous approach. The horizontal line imposed on the curve is the threshold.

exhaustive search still introduces a large number of false alarms. Figure 3.5 suggests that we should take advantage of the fact that audio markers can play a significant role in discarding many unnecessary candidates. In the next chapter, we will extend this discussion by adding the detection of visual markers.

3.3   Visual Marker Detection

3.3.1   MOTIVATION

When using low-level image features like color histogram, texture, and structure features to extract highlights (such as an attack on the soccer goal) from a 2- or 3-h long sports video, one faces the problem of an unacceptably large number of false alarms and misses. As mentioned in Chapter 2, those low-level image features are suitable for shot detection for scripted content, but they are not suitable for unscripted content to extract highlights. This is because shot detection is known to be unreliable for unscripted content. For example, in soccer video, visual features are so similar over a long period of time that almost all the frames within it may be grouped as a single shot. However, there might be multiple semantic units within the same period, such as attacks on the goal, counter attacks in the midfield, and so on. Hence, the gap between these low-level features and the semantic-level concepts is too large for these low-level features to bridge. This motivates us to look beyond low-level features. Instead, we want to detect highlights-related visual objects.

3.3.2   CHOICE OF VISUAL MARKERS

3.3.2.1   The Squatting Baseball Catcher

We make use of the following observation of the typical baseball video content: at the beginning of the baseball pitch, the camera angle is such that the catcher almost always faces the TV viewers, and we almost always see the frontal, upright view of the catcher squatting to catch the ball. Hence, there is a typical upright frontal view of the catcher in a squatting position waiting for the pitcher to pitch the ball. Some examples are shown in Figure 3.8. Although the umpire, the batter, and the pitcher are most of the time also in view, their poses and positions vary much more than those

Figure 3.8 Some examples of the typical view of the squatting baseball catcher.

of the catcher. For example, the batter can be on the right or left side of the catcher, or the umpire might be occluded by the catcher to different degrees. Robust identification of those video frames containing the catcher can bring us to the vicinity of the highlights. This motivates us to develop robust baseball catcher detection algorithms.

3.3.2.2   The Two Soccer Goalposts

We make use of the following two observations from soccer videos:

- When most of the interesting plays such as goals, corner kicks, or penalty kicks take place, the goalpost is almost always in view. Hence, detection of the goalpost can bring us to the vicinity of these interesting plays with high accuracy.
- There are two main views of the goalpost that we need to detect. To illustrate, we show a typical camera setup for broadcasting soccer games in Figure 3.9. The cameras are usually positioned in the center of the two sides of the field. The camera operators pan the camera in order to go back and forth between two halves of the field and zoom to focus on special targets. Since the distance between the camera and either of the two goalposts is relatively much larger than the size of the goalpost itself, little change occurs in the pose of the goalpost during the entire game, irrespective of the camera pan or zoom. These two typical views are also shown in Figure 3.9. Some of the example images are shown in Figures 3.10 and 3.11.

Robust identification of those video frames containing either of the two goalposts can bring us to the vicinity of soccer highlights. Because of the

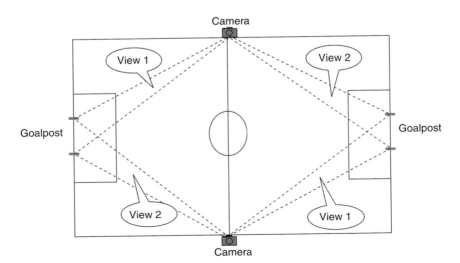

Figure 3.9   A typical video camera setup for live soccer broadcast.

Figure 3.10 Some examples of the typical view of the first view of the goalpost.

difference in appearance of these two views, we need to have two detectors that are responsible for detecting these two views respectively.

3.3.2.3 The Golfer Bending to Hit the Ball

It is harder to find visual markers in golf than in a baseball or soccer video. Two of the most appealing choices, the golf club and the golf

Figure 3.11   Some examples of the typical view of the second view of the goalpost.

ball, are unfortunately not easy to detect with high accuracy. For the golf club, key difficulties are different lighting conditions, different orientations of the club, and motion blur. For the golf ball, due to its small size, motion blur, and occlusion, color-based or shape-based approaches do not work well.

Figure 3.12   Some examples of the first view of the golfer.

We instead focus on three principal poses of the golfer. The first is the frontal (or nearly frontal) view of the golfer; the second is the side view with the golfer bending to the right; and the third is the opposite side view with the golfer bending to the left. Figures 3.12, 3.13, and 3.14 show several examples of these three views.

Figure 3.13   Some examples of the second view of the golfer.

Note that these three views are not enough to cover all the possible poses of the golfer. Although we have chosen a specific gesture of the golfer (body bending with two hands together prepared to hit the ball), we do not have much control of the poses. The choice of these three views is a compromise between detection accuracy and speed, which might be affected by too many views.

Figure 3.14 Some examples of the third view of the golfer.

We use an object detection algorithm to detect these six views (one for baseball, two for soccer, and three for golf) and an audio classification algorithm as shown in Section 3.2 to detect long and contiguous applause/cheering sounds for these three sports. A high detection rate of these key objects can eliminate most of the video content that is not in the vicinity of the interesting segments and bring us to the vicinity of the remarkable events in these three kinds of sports videos.

3.3.3 ROBUST REAL-TIME OBJECT DETECTION ALGORITHM

The object detection algorithm that we use is based on Viola and Jones's approach [46]. In the following, we first give an overview of the algorithm, we then explain why the algorithm achieves very fast detection.

Viola and Jones use an image representation called the "Integral Image," which allows the features used by the detector to be computed very quickly. Four kinds of features are listed in Figure 3.15.

Because these rectangle features can be of different sizes and at different locations with respect to the upper-left corner of the detection windows (e.g., of size 24 × 24), there are a large number of features. They then use a learning algorithm, based on AdaBoost, which selects a small number of critical rectangle features and yields extremely efficient classifiers. AdaBoost is an adaptive algorithm to boost a sequence of weak classifiers by dynamically changing the weights associated with the examples based on the errors learned previously so that more attention will be paid to the wrongly classified examples. The pseudo-code of AdaBoost is shown in Figure 3.16.

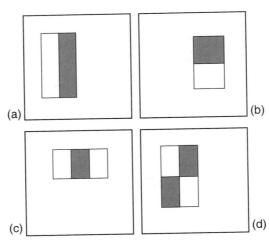

(a) (b) (c) (d)

Figure 3.15 Example rectangle features shown relative to the enclosing detection window. The sum of the pixels that lie within the white rectangles are subtracted from the sum of pixels in the gray rectangles. Two-rectangle features are shown in (a) and (b). Figure (c) shows a three-rectangle feature, and (d) shows a four-rectangle feature.

AdaBoost
Inputs
Training examples $(\mathbf{x}_1, y_1), \cdots, (\mathbf{x}_n, y_n)$ with $y_i \in \{0, 1\}$ and the number of iterations T.
Initialize weights $w_{1,i} = \frac{1}{2m}, \frac{1}{2l}$ for $y_i = 0, 1$ respectively, where m and l are the number of negative and positive examples respectively, with $l + m = n$.
Train Weak Classifiers
for $t = 1, \cdots, T$

(1) **Normalize** the weights, $w_{t,i} = \frac{w_{t,i}}{\sum_{j=1}^{n} w_{t,j}}$ so that w_t is a probability distribution.
(2) For each feature j, train a weak classifier h_j. The error is evaluated with respect to w_t, $\epsilon_j = \sum_i w_{t,i} |h_j(x_i) - y_i|$.
(3) Choose the best weak classifier, h_t, with the lowest error ϵ_t.
(4) **Update** the weights: $w_{t+1,i} = w_{t,i} \beta_t^{1-e_i}$ where $\beta_t = \frac{\epsilon_t}{1-\epsilon_t}$ and $e_i = 0$ if x_i is classified correctly, $e_i = 1$ otherwise.

Output
The final strong classifier is: $h(x) = sign(\sum_{t=1}^{T}(\alpha_t(h_t(x) - \frac{1}{2})))$ where $\alpha_t = \log \frac{1}{\beta_t}$.

Figure 3.16 The AdaBoost algorithm.

Furthermore, Viola and Jones use a method for combining classifiers in a cascade that allows the user to quickly discard background regions of the image while spending more computation time on promising object-like regions.

Two features have made Viola and Jones's algorithm very fast in detecting objects, such as frontal faces. First, there is no need to construct the so-called image pyramids that are necessary in many other face-detection algorithms to support an exhaustive search for faces of different sizes and at different locations. This is because the features (i.e., the difference between, e.g., the rectangle region) can be scaled proportionally to the size of the detection window. This difference can always be calculated in constant time (e.g., three additions) no matter how large the detection window is. By varying the size of the detection window, an exhaustive search equivalent to the one using "image pyramids" can be achieved in a significantly shorter time. Second, the "cascade" structure reflects the fact that within any single image an overwhelming majority of subwindows are negative.

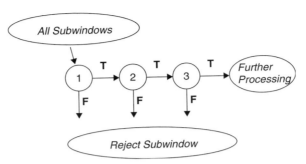

Figure 3.17 Schematic depiction of the detection cascade. A series of classifiers are applied to every subwindow. The initial classifier eliminates a large number of negative examples with very little processing. Subsequent layers eliminate additional negatives but require additional computation. After several stages of processing, the number of subwindows has been reduced radically. Further processing can take any form, such as additional stages of the cascade or an alternative detection system.

As such, the cascade attempts to reject as many negatives as possible at the earliest stage possible. While a positive instance will trigger the evaluation of every classifier in the cascade, this is an exceedingly rare event. This is illustrated in Figure 3.17.

3.3.4 RESULTS OF BASEBALL CATCHER DETECTION

3.3.4.1 Training of the Catcher Model

We have collected 1464 negative images (of size 352 pixels × 240 pixels), of which 240 are from several baseball games. These 240 images cover a large variety of the baseball playground, stadium, and audience views. The remaining 1224 images are taken from image collections that are not related to baseball at all. None of these images contains a baseball catcher. These images provide tens of millions of negative examples (of size 24 pixels × 24 pixels) for the learning process.

As for the positive examples, we have cropped 934 catcher images (also of size 24 pixels × 24 pixels) from three different baseball games. We have tried to include catchers wearing different jackets and clothes of different color. For each catcher image, we manually create nine deformed versions, namely, scaling it by a factor of 1.0, 0.9, and 0.8 and adding uniform noise of three different magnitudes. In the end, for every catcher image, we have produced nine images as positive examples. Hence, the total positive examples is 8406.

3.3.4.2 Interpretation of the Learned Features

Like the first few features for face detection [46], the initial rectangle features selected by AdaBoost based on the training baseball catcher images and negative examples are also meaningful and easily interpreted. We show the first three such features in Figure 3.18.

The first feature looks at the difference between "the sum of part of the leg region and the chest region" and "the sum of the hand region and the background region below the butt." The second feature reflects the difference between the chest region and the sum of the two arm regions. The third feature tells the difference between the very bottom portion of the image that has only the play field region (grass, soil, etc.) and the region right above it that has the player's feet.

3.3.4.3 Detection Results

We have applied the learned catcher models to detect the catchers from all the video frames in another 3-h long baseball game (Atlanta-Boston, Major League Baseball, 2002). If a detection is declared for a video frame, a binary number "1" is assigned to this frame. Otherwise, "0" is assigned. We have used the following technique to eliminate some false alarms in video marker detections: for every frame, we look at a range of frames corresponding to 1 s (starting from 14 frames before the current frame to 14 frames after the current frame). If the number of frames that have a

Figure 3.18 The first few weak classifiers learned by the AdaBoost algorithm for the catcher model.

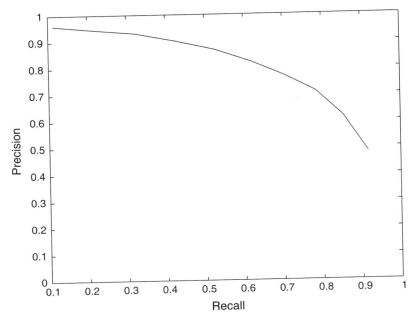

Figure 3.19   The precision-recall curve of baseball catcher detection.

detection declared is above a threshold, then we declare this frame as a frame that has detection. Otherwise, we declare this frame as a false positive. By varying this threshold (a percentage of the total number of frames in the range, in this case, 29), we can compare the number of detections with those in the ground truth set (marked by human viewers). We show the precision-recall curve for baseball catcher detection in Figure 3.19. We have achieved, for example, around 80% precision for a recall of 70%. More detailed results can be found in Table 3.6.

3.3.5   RESULTS OF SOCCER GOALPOST DETECTION

3.3.5.1   Intensity Normalization of the Goalposts

For each of the two goalpost views, we have followed the same procedures as those for catcher detection, including collecting positive examples and negative examples, training the models, and so forth. But we have observed that, unlike the learned catcher model that does well on the unseen test baseball game, the learned goalpost models do not work well at all on unseen

Table 3.6   **Precision-recall values for catcher detection.**

Threshold	Precision	Recall	Threshold	Precision	Recall
0.1	0.480	0.917	0.2	0.616	0.853
0.3	0.709	0.784	0.4	0.769	0.704
0.5	0.832	0.619	0.6	0.867	0.528
0.7	0.901	0.428	0.8	0.930	0.323
0.9	0.947	0.205	1.0	0.960	0.113

test soccer games. The "generalization" power of the goalpost models is much more limited than that of the catcher model. Our solution to this problem is through a preprocessing step—that is, intensity normalization to make use of the fact that the goalposts appear white, which occupies a very limited subspace of the color space. The details are as follows: for all the positive training images that have the goalposts and all the negative images, "quantize" the red, green, and blue (r, g, b) values of the image pixels according to the following formula:

$$(r, g, b) = \begin{cases} (255, 255, 255) & \text{if } r \geq 128 \ \& \ g \geq 128 \ \& \ b \geq 128 \\ (0, 0, 0) & \text{otherwise} \end{cases}$$

and apply the same formula to the video frames in the test soccer video. Note that the formula uses a "greedy" normalization scheme to bring different white pixels from the training images and the test images to the same value if they satisfy the condition. In the following description, this preprocessing step is taken for all the images. We show this preprocessing step in Figure 3.20.

3.3.5.2   Training of the Goalpost Models

For negative examples, we use the same set of images that we have used for the baseball catcher detection. As for the positive examples, for view 1, we have cropped 901 goalpost images (of size 24×18) from two different soccer games; for view 2, we have cropped 868 goalpost images (also of size 24×18) from the same two different soccer games. We have tried to include goalposts of different sizes and different lighting conditions. All these images first go through the preprocessing step described earlier.

Figure 3.20 Two examples of the preprocessing step for the two soccer goalpost views. The thumbnail images (bottom row) are taken from the "whitened" images (middle row).

For each goalpost image, we manually create nine deformed versions by adding uniform noise of nine different magnitudes. In the end, for every goalpost image, we have produced nine images as positive examples. Hence, the total positive examples is 8109 for view 1 and 7812 for view 2, respectively.

3.3.5.3   Interpretation of the Learned Features

Like the first few features for face detection [46], the initial rectangle features selected by AdaBoost based on the training goalpost images and negative examples are also meaningful and easily interpreted. We show the first three such features for view 1 in Figure 3.21 and three features for view 2 in Figure 3.22.

Each of the first two features accounts for the difference between a vertical bar region and its neighboring area. The third feature reflects the difference between the horizontal bar of the goalpost and its surrounding

Figure 3.21 The first few weak classifiers learned by the AdaBoost algorithm for the first view of the goalpost model.

Figure 3.22 The first few weak classifiers learned by the AdaBoost algorithm for the second view of the goalpost model.

area (although it appears tilted due to camera angles). Our interpretation is that these three features lock on to the edges of the goalposts due to a change of intensity.

3.3.5.4 Detection Results

We have applied the learned goalpost models to detect the goalposts from all the video frames in another 2-h long soccer game (Brazil-Germany, World Cup Soccer, 2002). If a detection is declared for a video frame, a binary number "1" is assigned to this frame. Otherwise, "0" is assigned. We have also used the same false alarm elimination scheme as the one for baseball catchers. We show the precision-recall curve for goalpost detection in Figure 3.23. We have achieved, for example, around 50% precision for a recall of 52% for view 1. More detailed results can be found in Table 3.7.

Comparing the goalpost detection results with the catcher detection results, we can see that goalpost detection does not achieve as high an accuracy as catcher detection using the AdaBoost-based algorithm. One reason might be that goalposts are more sensitive to motion blur than are baseball catchers. The localized features learned by the AdaBoost algorithm are not able to detect many frames in a contiguous sequence of goalpost frames where there are blurred goalposts due to camera motion.

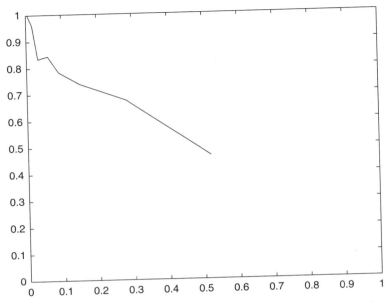

Figure 3.23 The precision-recall curve of goalpost detection for view 1.

Table 3.7 **Precision-recall values for goalpost detection for view 1.**

Threshold	Precision	Recall	Threshold	Precision	Recall
0.1	0.464	0.521	0.2	0.676	0.281
0.3	0.738	0.150	0.4	0.784	0.09
0.5	0.844	0.06	0.6	0.834	0.03
0.7	0.961	0.02	0.8	1.000	0.01

3.3.6 RESULTS OF GOLFER DETECTION

3.3.6.1 Training of the Golfer Models

For negative examples, we use the same set of images that we have used for the baseball catcher detection. As for the positive examples, for each of the three views, we have cropped 900 golfer images (of size 18×36) from several different golf games. We have included golfer images of different sizes, and different lighting conditions. For each golfer image, we manually create nine deformed versions by adding uniform noise of nine different magnitudes. In the end, for every goalpost image, we have produced nine images as positive examples. Hence, the total number of positive examples is 8100 for each of the three views.

3.3.6.2 Interpretation of the Learned Features

Like those for the baseball catcher and the two soccer goalposts, the initial rectangle features selected by AdaBoost based on the training golfer images and negative examples are also meaningful and easily interpreted. We show the first two such features for view 1 in Figure 3.24 and two features for view 2 in Figure 3.25.

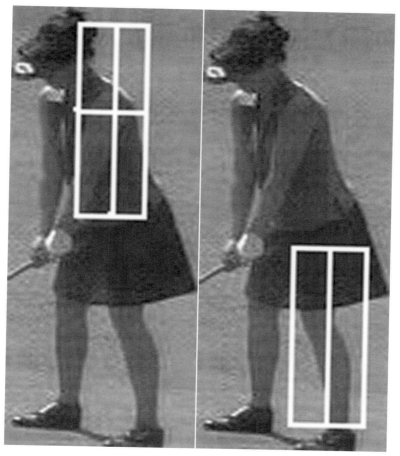

Figure 3.24 The first few weak classifiers learned by the AdaBoost algorithm for the first view of the golfer model.

Figure 3.25   The first few weak classifiers learned by the AdaBoost algorithm for the second view of the golfer model.

Although we still can interpret the features found in Figures 3.24 and 3.25, these interpretations are less intuitive than those for the catcher and the goalposts. (For view 1, the first feature emphasizes the difference between "the sum of the neck region and the waist region" and "the sum between the arm region and the background region behind the neck"; the second feature looks at the difference between the left- and right-hand side of the leg region. For view 2, the first feature shows the difference between the "sum of the head region and the hand region" and "twice the area of the chest region"; the second feature focuses on the butt region and the background region that is to the left of the butt.")

With these learned features, we have not been able to achieve a high detection rate with low false alarm rate on a different golf game (British Open, 2002). However, since there are so many different views of the golfer, it is difficult have a robust detector. Because golf audio is exceptionally easy to classify compared to baseball and soccer audio, we choose instead to rely on the audio cues in the golf video for highlights extraction. Hence, we use the AdaBoost-based visual object detection algorithm to detect the baseball catchers and the two soccer goalposts and the MDL-GMMs-based audio classification algorithm in Section 3.2 to detect long and contiguous applause in golf.

3.4 Finer-Resolution Highlights Extraction

3.4.1 AUDIO-VISUAL MARKER ASSOCIATION

Ideally each visual marker can be associated with one and only one audio marker and vice versa. Thus, they make a pair of audio-visual markers indicating the occurrence of a highlight event in their vicinity. But since many pairs might be wrongly grouped due to false detections and misses, some postprocessing is needed to keep the error to a minimum. We perform the following for associating an audio marker with a video marker.

- If a contiguous sequence of visual markers overlaps with a contiguous sequence of audio markers by a large margin (e.g., the percentage of overlapping is greater than 50%), then we form a "highlight" segment spanning from the beginning of the visual marker sequence to the end of the audio-visual marker sequence.
- Otherwise, we associate a visual marker sequence with the nearest audio marker sequence that follows it if the duration between the two is less than a duration threshold (e.g., the minimum duration of a set of training "highlights" clips from baseball games).

3.4.2 FINER-RESOLUTION HIGHLIGHTS CLASSIFICATION

Highlight candidates, delimited by the audio markers and visual markers, are quite diverse. For example, golf swings and putts share the same audio markers (audience applause and cheering) and visual markers (golfers bending to hit the ball). Both of these golf highlight events can be found by the aforementioned audio-visual marker detection-based method. To support the task of retrieving finer events such as "golf swings only" or "golf putts only," we have developed techniques that model these events using low-level audio-visual features. Furthermore, some of these candidates might not be true highlights. We eliminate these false candidates using a finer-level highlight classification method. For example, for golf, we build models for golf swings, golf putts, and nonhighlights (neither swings nor putts) and use these models for highlights classification (swings or putts) and verification (highlights or nonhighlights).

In this chapter, we propose three methods for this task. The first method makes use of the observation that finer-resolution highlight events have different "global" color characteristics such as "no large change over time" or "a large change over time." The second method models these characteristics using HMMs trained using example clips. The third method uses the additional audio information. It uses CHMM, a joint audio-visual fusion technique for the task. CHMMs, which have been applied to audio-visual speech recognition [47] and complex action recognition [48], to our knowledge have not been applied to problems in this domain.

3.4.3   *METHOD 1: CLUSTERING*

As an example, let us look at finer-level highlight classification for a baseball game using low-level color features. The diverse baseball highlight candidates found after the audio-visual marker negotiation step are further separated using the techniques described here. For baseball, there are two major categories of highlight candidates, the first being "balls or strikes" in which the batter does not hit the ball, the second being "ball-hits" in which the batter hits the ball to the field or audience. These two categories have different color patterns. In the first category, the camera is fixed at the pitch scene, so the variance of color distribution over time is low. In the second category, the camera first shoots the pitch scene, and then it follows the ball to the field or the audience. The variance of color distribution over time is therefore high. Figure 3.26 shows an example.

We extract the 16-bin color histogram using the hue component in the hue-saturation-value (HSV) color space from every video frame of each of the highlight candidate video clips. So every highlight candidate is represented by a matrix of size $L \times 16$, where L is its number of video frames. Let us denote this matrix as the "color histogram matrix." We use the following algorithm to do the finer-resolution highlights classification:

- For every color histogram matrix, calculate the "clip-level" mean vector (of length 16) and the "clip-level" standard deviation (STD) vector (also of length 16) over its rows.

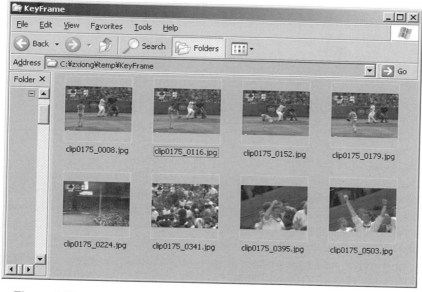

Figure 3.26 An example of the change of color characteristics in a baseball hit.

- Cluster all the highlight candidate video clips based on their "clip-level" STD vectors into two clusters. The clustering algorithm we have used is the k-means algorithm.
- For each of the clusters, calculate the "cluster-level" mean vector (of length 16) and the "cluster-level" STD vector (also of length 16) over the rows of all of the color histogram matrices within the cluster.
- If the value at any color bin of the "clip-level" mean vector is outside the 3σ range of the "cluster-level" mean vector where σ is the STD of the "cluster-level" STD vector at the corresponding color bin, remove it from the highlight candidate list.

When testing on a 3-h long baseball game, of all the 205 highlight candidates, we have removed 32 using this algorithm. These 32 clips have been further confirmed to be false alarms by human viewers (see Table 3.8).

3.4.4 METHOD 2: COLOR/MOTION MODELING USING HMMs

Although some of the false alarm highlights returned by audio classification have long contiguous applause segments, they do not contain real

Table 3.8   **Results before and after the finer-resolution highlights classification.**

Video Length		540 Min
Method	A-V Negotiation	Finer-Resolution Classification
Number of highlight candidates	205	173
Number of false alarms	50	18
Highlights length	29.5 minutes	25.8 minutes

highlight actions in play. For example, when a player is introduced to the audience, applause abounds. We have noticed that the visual patterns in such segments are quite different from those in highlight segments such as "putt" or "swing" in golf. These visual patterns include the change of dominant color and motion intensity. In putt segments, the player stands in the middle of the golf field that is usually green, which is the dominant color in the golf video. In contrast, when the announcer introduces a player to the audience, the camera is usually focused on the announcer, so there is not much green color of the golf field. In swing segments, the golf ball goes from the ground up, flies against the sky, and comes down to the ground. In the process, there is a change of color from the color of the sky to the color of the playing field. Note there are two dominant colors in swing segments. Also, since the camera follows the ups and downs of the golf ball, there is the characteristic pan and zoom, both of which may be captured by the motion intensity features.

3.4.4.1 Modeling Highlights by Color Using HMM

We divide the 50 highlight segments we collected from a golf video into two categories: 18 putt and 32 swing video sequences. We use them to train a putt and a swing HMM, respectively, and test on another 95 highlight segments we collected from another golf video. Since we have the ground truth of these 95 highlight segments (i.e., whether they are putt or swing), we use the classification accuracy on these 95 highlight segments to guide us in search of the good color features.

First, we use the average hue value of all the pixels in an image frame as the frame feature. The color space here is the HSV space. For each of the 18 putt training sequences, we model the average hue values of all the video frames using a three-state HMM. In the HMM, the observations (i.e., the average hue values) are modeled using a three-mixture GMM.

We model the swing HMM in a similar way. When we use the learned putt and swing HMMs to classify the 95 highlight segments from another golf video, the classification accuracy is quite low, ~60% on average over many runs of experiments.

Next, noticing that the range of the average hue values are quite different between the segments from the two different golf videos, we use the following scaling scheme to make them comparable to each other: for each frame, divide its average hue value by the maximum of the average hue values of all the frames in each sequence. With proper scaling by a constant factor (e.g., 1000), we are able to improve the classification accuracy from ~60% to ~90%.

In Figures 3.27 through 3.29, we have plotted these average hue values of all the frames for the 18 putt and 32 swing video sequences for training and the 95 video sequences for testing, respectively.

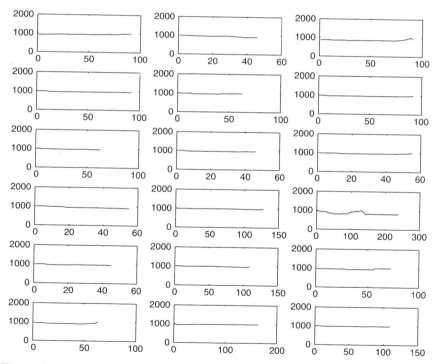

Figure 3.27 The scaled version of each video frame's average hue value over time for the 18 training "putt" sequences. The scaling factor is 1000/MAX(·). X-axis: video frames; Y-axis: scaled average hue values.

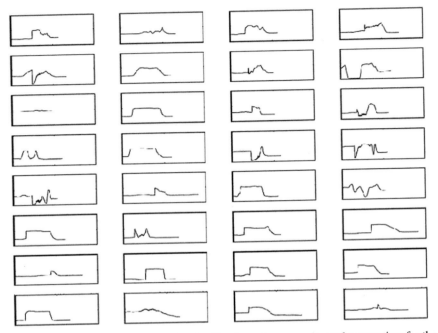

Figure 3.28   The scaled version of each video frame's average hue value over time for the 32 training "swing" sequences. The scaling factor is 1000/MAX(·). X-axis: video frames; Y-axis: scaled average hue values.

Note that the putt color pattern in Figure 3.27 is quite different from that of the swing color pattern in Figure 3.28. This difference is also shown in the color pattern of the test sequences in Figure 3.29.

3.4.4.2   Further Verification by Dominant Color

The scaling scheme mentioned earlier does not perform well in differentiating "uniform" green color for putt from uniform color of an announcer's clothes in a close video shot. To resolve this confusion, we learn the dominant green color [6] from those candidate highlight segments indicated by the GMM-MDL audio classification. The grass color of the golf field is the dominant color in this domain, since a televised golf game is bound to show the golf field most of the time in order to correctly convey the game status. The appearance of the grass color, however, ranges from dark green to yellowish green or olive, depending on the field condition and capturing device. Despite these factors, we have observed that within one game,

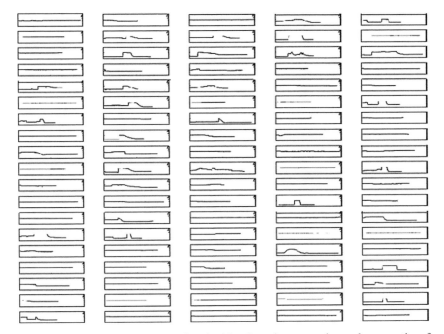

Figure 3.29 The scaled version of each video frame's average hue value over time for the 95 test sequences.

the hue value in the HSV color space is relatively stable despite lighting variations, hence learning the hue value would yield a good definition of dominant color. The dominant color is learned adaptively from those candidate highlight segments using the following cumulative statistics: average the hue values of the pixels from all the video frames of those segments to be the center of the dominant color range; use twice the variance of the hue values over all the frames as the bandwidth of the dominant color range.

3.4.4.3 Modeling Highlights by Motion Using HMM

Motion intensity m is computed as the average magnitude of the effective motion vectors in a frame:

$$m = \frac{1}{|\Phi|} \sum_{\Phi} \sqrt{v_x^2 + v_y^2} \qquad (3.4.1)$$

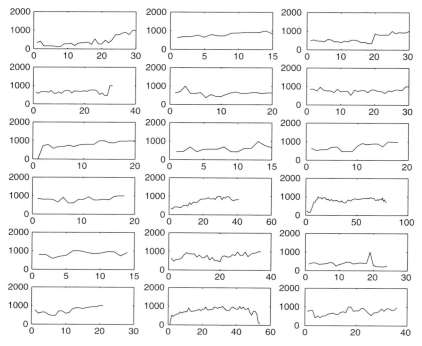

Figure 3.30 The scaled version of each video frame's average motion intensity value over time for the 18 training "putt" sequences. The scaling factor is 1000/MAX($\cdot$). X-axis: video P-frames; Y-axis: scaled average motion intensity values.

where $\Phi = \{\text{intercoded macroblocks}\}$ and $v = (v_x, v_y)$ is the motion vector for each macroblock. This measure of motion intensity gives an estimate of the gross motion in the whole frame, including object and camera motion. Moreover, motion intensity carries complementary information to the color feature, and it often indicates the semantics within a particular shot. For instance, a wide shot with high motion intensity often results from player motion and camera pan during a play, while a static wide shot usually occurs when the game has come to a pause. With the same scaling scheme as the one for color, we are able to achieve a classification accuracy of $\sim$80% on the same 95 test sequences.

We have plotted the average motion intensity values of all the frames of all the sequences in Figures 3.30 through 3.32 for the 18 putt and 32 swing video sequences for training and the 95 video sequences for testing, respectively.

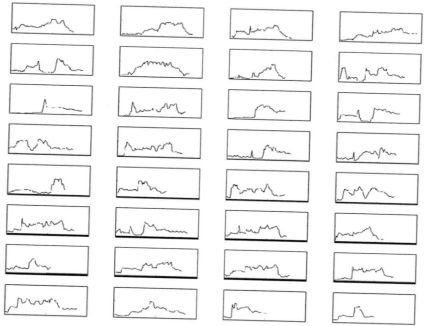

Figure 3.31   The scaled version of each video frame's average motion intensity value over time for the 32 training "swing" sequences. The scaling factor is 1000/MAX($\cdot$). X-axis: video P-frames; Y-axis: scaled average motion intensity values.

3.4.4.4   Proposed Audio-Visual Modeling Algorithm

Based on these observations, we model the color pattern and motion pattern using HMMs. We learn a putt HMM and a swing HMM of the color features. We also learn a putt HMM and a swing HMM of the motion intensity features.

Our algorithm can be summarized as follows:

(1) Analyze the audio to locate contiguous applause segments.

- Perform silence detection.
- For nonsilent segments, run the GMM-MDL classification algorithm using the trained GMM-MDL models.

(2) Sort those contiguous applause segments based on the applause length.

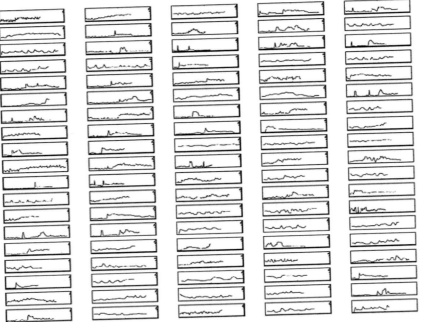

Figure 3.32 The scaled version of each video frame's average motion intensity value over time for the 95 test sequences. The scaling factor is 1000/MAX(·). X-axis: video P-frames; Y-axis: scaled average motion intensity values.

(3) Analyze the video to verify whether or not those applause segments follow the correct color and motion pattern.

- Take a certain number of video frames before the onset of each of the applause segments to estimate the dominant color range.
- For a certain number of video frames before the onset of each of the applause segments, run the putt or swing HMM of the color features.
- If it is classified as putt, then verify that its dominant color is in the estimated dominant color range. If the color does not fit, then declare it as a false alarm and eliminate its candidacy.
- If it is classified as swing, then run the putt or swing HMM of the motion intensity features; if it is classified again as swing we say it is swing; otherwise, declare it as a false alarm and eliminate its candidacy.

3.4.4.5 Experimental Results, Observations, and Comparisons

We further analyze the extracted highlights that are based on those segments in Table 3.5. For each contiguous applause segment, we extract a
certain number of video frames before the onset of the detected applause.
The number of the video frames is proportional to the average video
frames of those putt or swing sequences in the training set. For these video
frames, we verify whether they are of putts or swings using the proposed
algorithm. To compare with the precision-recall curve on the left-hand side
of Figure 3.5, we plot two more precision-recall curves to compare with
the "GMM-MDL audio" approach, one being the "GMM-MDL audio classification + color HMM" approach and the other being the "GMM-MDL
audio classification + color HMM + motion intensity HMM" approach
in Figure 3.33.

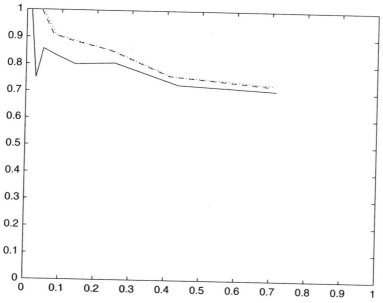

Figure 3.33 Comparison results of three different modeling approaches in terms of
precision-recall curves. Solid line: audio modeling alone; dashed line: audio + dominant
color modeling; dotted line: audio + dominant color + motion modeling. X-axis: recall;
Y-axis: precision.

The following observations can be made from the precision-recall comparison in Figure 3.33:

- Both the dashed curve and the dotted curve representing "audio modeling + visual modeling" show better precision-recall figures. By carefully examining where the improvement comes from, we notice that the application of color and motion modeling has eliminated such false alarms as those involving the announcer or video sequences followed by non-applause audio. By jointly modeling audio and visual features for sports highlights, we have been able to eliminate these two kinds of false alarms: wrong video pattern followed by applause and video pattern followed by nonapplause.

- Between the dashed curve and the dotted curve, although the introduction of additional motion intensity modeling improves performance over the "audio modeling + color modeling," the improvement is only marginal.

3.4.5 METHOD 3: AUDIO-VISUAL MODELING USING CHMMs

A coupled HMM is a collection of HMMs, one for each data stream, that are coupled through the transitional probabilities from the state nodes at $t - 1$ of all the HMMs to each of the state nodes at t of all the HMMs. Since we are focused on only two streams—the audio and visual stream—we only describe the coupling of two HMMs. Figure 3.34 illustrates a two-stream discrete observation coupled HMM (DCHMM).

3.4.5.1 Audio Label Generation

In the audio domain, there are common events relating to highlights across different sports. After an interesting golf hit or baseball hit or an exciting soccer attack, the audience shows appreciation by applauding or even cheering loudly. The duration of applause or cheering is usually longer when the play is more interesting (e.g., a home run in baseball). There are also common events relating to uninteresting segments in sports TV broadcasting (e.g., TV commercials that are mainly composed of music or speech with music segments).

We therefore mainly focus on building models for the following seven classes of audio signals: applause, ball-hit, cheering, music, female speech, male speech, and speech with music. The motivation for choosing these

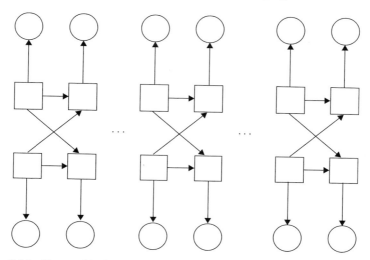

Figure 3.34 The graphical model structure of the DCHMM. The two rows of squares are the coupled hidden state nodes. The circles are the observation nodes.

Table 3.9 **Audio labels and class names.**

Audio Label	*Its Meaning*	*Audio Label*	*Its Meaning*
1	Silence	5	Music
2	Applause	6	Female speech
3	Ball-hit	7	Male speech
4	Cheering	8	Speech with music

seven classes is that the former three are indicators of "interesting" sports events that are directly related to highlights, and correct recognition of the latter four can filter out most of the uninteresting segments.

The models we build for these seven classes are the GMMs. Apart from the entire sports audio sound track, we have collected a separate set of audio data that covers all seven classes. We train the GMMs from 90% of these data and use the classification accuracy on the remaining 10% to compare and select the best from different model parameters. More detailed experimental results are reported in Section 3.4.6. We list these classes together with the silence class in Table 3.9.

3.4.5.2 Video Label Generation

We use a modified version of the MPEG-7 motion activity descriptor to generate video labels. The MPEG-7 motion activity descriptor captures the intuitive notion of "intensity of action" or "pace of action" in a video segment [49]. It is extracted by quantizing the variance of the magnitude of the motion vectors from those video frames between two neighboring P-frames to one of five possible levels: very low, low, medium, high, and very high. Since Peker et al. [49] have shown that the average motion vector magnitude also works well with lower computational complexity, we adopt this scheme and quantize the average of the magnitudes of motion vectors from those video frames between two neighboring P-frames to one of four levels: very low, low, medium, and high. These labels are listed in Table 3.10.

3.4.5.3 Training Highlight and Non-highlight Models with CHMMs

We apply the CHMM-based audio-visual fusion to train the highlight models. We have collected both positive examples (video clips of golf swing and putt for golf; goal shooting, corner kick, or fast attack for soccer; etc.) and negative examples (video clips of commercials or golf swing without audience applause in golf; replays, scenes of commentator speaking, or midfield soccer plays in soccer; etc.). First, we extract audio and motion class labels from these training examples using the techniques described in the previous two subsections. Then we align the two sequences of labels by down-sampling the video labels to match the length of the audio label sequence for every video clip. Next, we carefully choose the number of states of the CHMMs by analyzing the semantic meaning of the labels corresponding to each state decoded by the Viterbi algorithm. Finally, we show

Table 3.10 **Video labels and class names.**

Video Label	Its Meaning
1	Very low motion
2	Low motion
3	Medium motion
4	High motion

that the chosen number of states also achieves better classification accuracy when tested on the test data. More details can be found in Section 3.4.6.

3.4.5.4   Extracting Highlights from Test Video Games

Since our task is to extract highlight segments from a several-hour-long sports video, we first need to determine where the candidates are—that is, from which video frame they start from and at which video frame they end. We use the audio classification results from the aforementioned seven-class audio GMMs. First, we use audio energy to detect silent segments. Then, we classify every second of nonsilence audio into one of the seven classes. We locate the contiguous applause segments in golf or cheering segments in soccer and then use an equal length of video before and after the onset of these segments as a candidate. Next, we "cut" these candidates from the video and extract audio and motion class labels to feed them to the highlight CHMM or nonhighlight CHMM. Then, we use the maximum likelihood criterion to decide whether the candidate is a highlight in golf or soccer. Finally, we filter those classified highlights that do not overlap in time with those of the ground truth set and declare the remaining candidates as true highlights.

3.4.6   EXPERIMENTAL RESULTS WITH DCHMM

3.4.6.1   Results on Audio Classification

We have collected more than 3 h of audio clips from TV broadcasting of golf, baseball, and soccer games. Each of them is hand-labeled into one of the seven classes as ground truth. We extract a hundred 12-dimensional MFCCs per second using a 25 μs window. We also add the first- and second-order time derivatives to the basic MFCC parameters in order to enhance performance. For more details on MFCC feature extraction see [44].

We use 90% of the collected audio clips for training GMMs and use the classification accuracy on the remaining 10% by these models to fine-tune the choice of the their parameters, such as number of mixtures. We give the best classification accuracy results in Table 3.11 using a mixture of 10 Gaussians for each of the audio classes using diagonal covariance matrices.

3.4.6.2   Results on Golf Highlights Extraction

We have collected 50 golf highlight video clips as positive training examples from a 2-h long golf game (Golf LPGA Tour, 2002). These video clips

Table 3.11 **Recognition matrix (or confusion matrix) on a 90%/10% training/testing split of a data set composed of seven classes: (1) applause, (2) ball-hit, (3) cheering, (4) music, (5) female speech, (6) male speech, and (7) speech with music. The results here are based on MFCCs and GMMs.**

	(1)	(2)	(3)	(4)	(5)	(6)	(7)
(1)	1.00	0	0	0	0	0	0
(2)	0	0.846	0	0	0.077	0.077	0
(3)	0	0	1	0	0	0	0
(4)	0.056	0	0	0.889	0.028	0	0.028
(5)	0	0	0	0	0.941	0.059	0
(6)	0	0	0	0	0	0.941	0.059
(7)	0	0	0	0	0.069	0.051	0.88

Average Recognition Rate: 92.8%

consist of nice golf swings that land the ball close to the target hole followed by applause from the audience. They also consist of good putts that lead to a lot of audience applause. We have also collected 50 video clips that are similar to but different in nature from those positive training examples as negative examples. These clips consist of golf swings without audience applause because the ball lands far away from the target hole, and TV commercial segments. We train both the golf highlight and nonhighlight models using the CHMM-based approach described in Section 3.4.5.

We then test the learned golf highlight and nonhighlight models on a different golf game (a 3-h long British Open 2002 game). We first locate those long, contiguous applause segments recognized by the audio GMMs. Of the 233 applause segments that are found, the longest one lasts 13 s, the shortest lasts 2 s (we discarded the sporadic 1-s applause segments because most of the true applause is more than 1-s long and most of these segments also contain many false classifications). We then extract audio and motion labels at the neighborhood of these candidate positions. To evaluate the highlight extraction accuracy, we have hand-marked the position of 95 highlight segments in the British Open 2002 game. These 95 highlight segments are used as the ground truth in evaluation.

In the following, we show how we fine-tune some of the model parameters in training the golf highlight models. To improve the capability of modeling the label sequences, we follow Rabiner's [50] description of refinement on the model (choice of the number of states, different codebook size, etc.) by segmenting each of the training label sequences into states and then studying the properties of the observed labels occurring in each state. Note that the states are decoded via the Viterbi algorithm in an unsupervised fashion (i.e., unsupervised HMM).

We first show the refinement on the number of states for both the "audio-alone" and the "video-alone" approach. With an appropriate number of states, the physical meaning of the model states can be easily interpreted. We then build the CHMM using these refined states. We next compare the results of these three different approaches (i.e., "audio-alone," "video-alone," and the CHMM-based approach).

3.4.6.3 The "Audio-Alone" Approach

The following are some sequences of classified audio labels on 50 highlight audio sound tracks used for training. Note that all these sequences have a contiguous "applause" label at the end:

$$1, 3, 3, 3, 7, 3, 3, 3, 2, 2, 2, 2$$

$$1, 3, 3, 2, 2, 2$$

$$3, 7, 7, 2, 2, 2, 2, 2, 2$$

$$\cdots$$

$$3, 3, 3, 5, 5, 2, 2, 2$$

$$3, 3, 3, 2, 2, 2, 3, 2$$

$$\cdots$$

In optimizing the number of states using the unsupervised HMM, we have found that a total number of two states give the most interpretable meaning of the states. One example of the learned state transition matrix is as follows:

$$\mathbf{A}^a = \begin{pmatrix} 0.9533 & 0.0467 \\ 0.2030 & 0.7970 \end{pmatrix} \tag{3.4.2}$$

Its corresponding observation matrix is

$$\mathbf{B}^a = \begin{pmatrix} 0 & 0.82 & 0.03 & 0 & 0.03 & 0 & 0.11 & 0.01 \\ 0.15 & 0.01 & 0.54 & 0.02 & 0.15 & 0.02 & 0.10 & 0 \end{pmatrix} \tag{3.4.3}$$

Table 3.12   **Number of observed labels in each state.**

S\O	1	2	3	4	5	6	7	8
1	0	185	9	0	5	0	31	2
2	36	0	130	6	38	5	20	0

When we segment each of the training label sequences into these two states, we notice that there is a clear difference of the observed labels associated with them. One state has the observation of "applause" most of the time, while the other state has all the other "nonapplause" labels. This can be seen from Equation (3.4.3) and Table 3.12.

Note that although the numbers in Equation (3.4.2), Equation (3.4.3), and Table 3.12 change when we rerun the unsupervised HMM due to the randomized initialization of the HMM parameters, the pattern stays the same. So it is reasonable to assume that of the two states, one state corresponds to the "applause" sound (the first state in the preceding example) and the other corresponds to the nonapplause sound (the second state).

When we split both the 50 positive and the negative clips into a 90% versus 10% training versus test set, we compare the test results on the five test clips using the models trained from the 45 training clips for different numbers of HMM states. We have done this split 10 times, every time a random split is chosen. The results in Table 3.13 show that the averaged classification accuracy is the best for the aforementioned two states.

Table 3.13   **Averaged classification accuracy on 10% of the total golf highlight and nonhighlight data using the audio HMMs.**

Number of States	Accuracy
1	90.9%
2	96.9%
3	94.1%
4	92.9%
5	91.5%

3.4.6.4 The "Video-Alone" Approach

The following are the corresponding sequences of video labels on the video of the same 50 highlights. Note the existence of a "low-high-low" motion evolution pattern:

$$2, 3, 3, 3, 4, 4, 2, 1, 2, 2, 2, 1$$
$$1, 1, 2, 3, 3, 4$$
$$1, 1, 3, 4, 2, 4, 4, 3, 3$$
$$\cdots$$
$$3, 3, 4, 3, 3, 2, 3, 3$$
$$4, 3, 4, 4, 4, 4, 3, 3$$
$$\cdots$$

In optimizing the number of states using the unsupervised HMM, we have found that a total number of two states gives the most interpretable meaning of the states. One example of the learned state transition matrix is shown here:

$$\mathbf{A}^v = \begin{pmatrix} 0.8372 & 0.1628 \\ 0.2370 & 0.7693 \end{pmatrix} \tag{3.4.4}$$

Its corresponding observation matrix is

$$\mathbf{B}^v = \begin{pmatrix} 0.0 & 0.05 & 0.57 & 0.38 \\ 0.29 & 0.70 & 0.01 & 0.00 \end{pmatrix} \tag{3.4.5}$$

When we segment each of the training label sequences into these two states, we notice that there is also a clear difference of the observed labels associated with them. One state has the observation of either "medium motion" or "high motion" most of the time, while the other state has "very low motion" or "low motion" labels. This can be seen from Equation (3.4.5) and Table 3.14.

Also note that although the numbers in Equation (3.4.4), Equation (3.4.5), and Table 3.14 change when we rerun the unsupervised HMM due to the randomized initialization of the HMM parameters, the pattern stays the same. So it is reasonable to assume that of the two states, one state corresponds to low motion (the second state in this example) and the other corresponds to high motion (the first state).

Table 3.14 **Number of observed labels in each state.**

S\O	1	2	3	4
1	0	14	148	96
2	61	146	0	0

Table 3.15 **Averaged classification accuracy on 10% of the total golf highlight and nonhighlight data using the video HMMs.**

Number of States	Accuracy
1	92.7%
2	95.8%
3	91.6%
4	85.1%
5	82.3%

The parallel experiments to those using audio HMM show the averaged classification accuracy in Table 3.15. We can see that with a choice of two states for the video HMM, the averaged classification accuracy on the 10% of the total training data from the models trained from the remaining 90% of the data gives best performance.

3.4.6.5   Results of the CHMM Approach

After refining the states in the previous two single-modality HMMs, we build the CHMM with two states for the audio HMM and two states for the video HMM and introduce the coupling between the states of these two models. When we repeat the unsupervised CHMM on both the audio and video labels, we still can interpret the learned coupled state transition matrix and observation matrix. In Equations (3.4.6) and (3.4.7), we show the coupled state transition matrix as an example.

We show the averaged classification accuracy in Table 3.16 again on the 90%/10% split of the training data, this time using CHMMs. We can see that with a choice of two states for the video HMM and a choice of two states for the audio HMM gives the best performance. We plot four PR curves on the test game in Figure 3.35. The solid-line curve is based on the assumption that all the candidates containing long, contiguous applause segments are highlights. The dashed-line curve is based on the audio HMM tested on those candidates segments. The dash-dot-line curve is based on the visual HMM tested on those candidates segments. The dotted-line curve is based on the joint CHMM modeling tested on those candidates. Note that for this golf game, the first and the second curve coincide with each other. Our inspection reveals that because the accuracy of applause classification is very high, the audio HMM does not eliminate any false alarms.

$$\mathbf{A_c^a} = Pr(S_a(t+1) = k | S_a(t) = i, S_v(t) = j)$$

	$k = 1$	$k = 2$
$(i = 1, j = 1)$	0.8178	0.1822
$(i = 1, j = 2)$	0.7958	0.2042
$(i = 2, j = 1)$	0.0810	0.9190
$(i = 2, j = 2)$	0.2093	0.7907

(3.4.6)

$$\mathbf{A_c^v} = Pr(S_v(t+1) = l | S_a(t) = i, S_v(t) = j)$$

	$l = 1$	$l = 2$
$(i = 1, j = 1)$	0.9302	0.0698
$(i = 1, j = 2)$	0.2045	0.7955
$(i = 2, j = 1)$	0.9157	0.0843
$(i = 2, j = 2)$	0.1594	0.8406

(3.4.7)

Table 3.16 Averaged classification accuracy on 10% of the total golf highlight and nonhighlight data using the CHMMs.

Number of States	Accuracy	Number of States	Accuracy	Number of States	Accuracy
$M = 1, N = 1$	91.7%	$M = 1, N = 2$	92.1%	$M = 1, N = 3$	91.6%
$M = 2, N = 1$	92.7%	$M = 2, N = 2$	96.8%	$M = 2, N = 3$	91.6%
$M = 3, N = 1$	89.7%	$M = 3, N = 2$	91.7%	$M = 3, N = 3$	91.6%

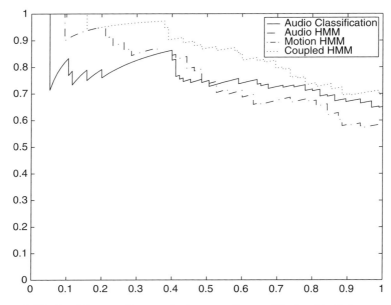

Figure 3.35 Precision-recall curves for the test golf game. The four highlights extraction methods compared here are (1) audio classification followed by long, contiguous applause selection; (2) HMM classification using the models trained from audio labels of the highlight and nonhighlight examples; (3) HMM classification using the models trained from video (motion) labels of the highlight and nonhighlight examples; and (4) coupled HMM classification using the models trained from both audio and video (motion) labels of the highlight and nonhighlight examples. X-axis: recall; Y-axis: precision. The first two curves coincide with each other (see text for detail).

Comparing the four PR curves in Figure 3.35, we can make the following observations:

(1) The CHMM-based approach shows higher precision values than either single modality HMM-based approach for all different recall values. This implies that the CHMM-based approach has a lower false alarm rate for a given recall rate. We believe the promising gain in performance is due to the coupling between the audio and visual modeling.

(2) All four approaches achieve more than a 60% precision rate for all different recall values. For this 3-h golf game, when the recall rate is 100% (i.e., all the highlights are extracted), the total length of the

extracted highlights (including true ones and false alarms) lasts less than 10 min even for the solid-line curve. This suggests that overall the highlight extraction results are satisfactory.

3.4.6.6 Results on Soccer Highlights Extraction

We have collected 50 soccer highlight video clips as positive training examples from a 2-h long TV broadcast soccer game (World Cup Japan versus Turkey, 2002). These video clips consist of soccer goal shots, fast attacks, and corner kick attacks that result in loud audience cheering. We have also collected 50 video clips as negative examples that consist of midfield back-and-forth scenes, commentator speaking, and TV commercials. We train both the soccer highlight and nonhighlight models using the CHMM-based approach described in Section 3.4.5.

We do the same segmentation of the observation labels based on the Viterbi-decoded state sequences in order to find coherent correspondence between state and its associated observed labels. For motion activity modeling, we find that two states still make the most interpretable segmentation, just like the one for the golf highlights. That is, when we segment each of the training label sequences into two states, we notice that there is a clear difference of the observed labels associated with them. On the other hand, we find that for audio modeling, three states instead of two result in the most interpretable segmentation: one for "applause," one for "cheering," and the third for "nonaudience-reaction sound." We also compared different combinations of the number of audio and video HMM states, and we put the results in Table 3.17. We can see that with this combination of three audio HMM states and two video HMM states, the averaged classification

Table 3.17   **Averaged classification accuracy on 10% of the total soccer highlight and nonhighlight data using CHMMs.**

Number of States	*Accuracy*	*Number of States*	*Accuracy*	*Number of States*	*Accuracy*
$M = 1, N = 1$	91.7%	$M = 1, N = 2$	92.1%	$M = 1, N = 3$	91.6%
$M = 2, N = 1$	92.7%	$M = 2, N = 2$	93.8%	$M = 2, N = 3$	91.6%
$M = 3, N = 1$	89.7%	$M = 3, N = 2$	95.7%	$M = 3, N = 3$	91.6%
$M = 4, N = 1$	89.7%	$M = 4, N = 2$	91.7%	$M = 4, N = 3$	91.6%
$M = 5, N = 1$	89.7%	$M = 5, N = 2$	91.7%	$M = 5, N = 3$	91.6%

accuracy on the 10% of the total training data from the models trained from the remaining 90% of the data gives the best performance.

We have tested the learned soccer highlight and nonhighlight models on a different golf game (the 2-h long World Cup final game, Brazil versus Germany, 2002). We first locate those long, contiguous cheering segments recognized by the audio GMMs. Of all 323 cheering segments that are found, the longest cheering lasts 25 s, the shortest lasts 2 s (we again discarded the sporadic 1-s cheering segments using the same reason as that for golf). We then extract audio and motion labels at the neighborhood of these candidate positions. To evaluate the highlight extraction accuracy, we have hand-marked the position of 43 highlight segments in the World Cup final game. These 43 highlight segments are used as the ground truth in evaluation.

Like Figure 3.35, we plot four PR curves on the test World Cup final game in Figure 3.36. Note there is a significant difference between these

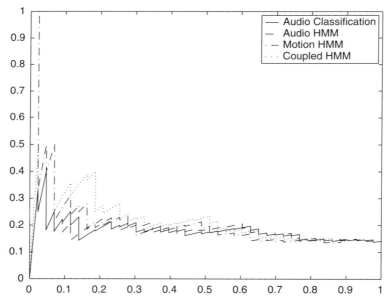

Figure 3.36 Precision-recall curves for the test soccer game. The four highlights extraction methods compared here are (1) audio classification followed by long, contiguous cheering selection; (2) HMM classification using the models trained from audio labels of the highlight and nonhighlight examples; (3) HMM classification using the models trained from video (motion) labels of the highlight and nonhighlight examples; and (4) coupled HMM classification using the models trained from both audio and video (motion) labels of the highlight and nonhighlight examples. X-axis: recall; Y-axis: precision.

two figures. For this soccer game, the first and the second curve do not coincide with each other any more. This is because the soccer audio sound track has a high level of background audience noise, which is easily confused with the real cheering due to interesting plays. We can see that audio HMM trained on highlight examples has eliminated some false alarms.

Comparing the four PR curves in Figure 3.36, we can make the following observations:

(1) The CHMM-based approach again shows higher precision values than either single modality HMM-based approach for almost all different recall values. This implies that the CHMM-based approach has a lower false alarm rate for a given recall rate. We believe the gain in performance is due to the coupling between the audio and visual modeling.

(2) Compared with the curves in Figure 3.35, the overall performance for the soccer game is worse than that for the golf game. For this 2-h soccer game, when the recall rate is 100% (i.e., all the highlights are extracted), the false alarm rate is around 70%, implying that there are many false alarms besides the truth highlights.

When we view the highlights using the interface in Figure 3.37 for the World Cup final game, we have the following observations:

(1) Many sharp peaks in Figure 3.37 that are classified as highlights indeed are false alarms. These false alarms do share the property with many of the highlights, for example, midfield fighting with high motion activity together with a high level of background sound that is classified as cheering. We believe more work needs to be done to differentiate these similar audio classes.

(2) Most of the very uninteresting events, such as commercials inserted between the sports content, can still be filtered out. This implies that most of the false alarms are those difficult-to-tell-apart elements of content.

We have presented three methods for finer-resolution highlight classification. What is common among these three methods is that they all operate after highlight candidates have been found, either by audio marker detection for golf or by joint audio-visual association for baseball and soccer. Their differences lie in complexity and supervision. The first method, although simple, involves parameter tuning, so it might not work well across different content. The last two methods increase in complexity using trained finer-resolution highlight models.

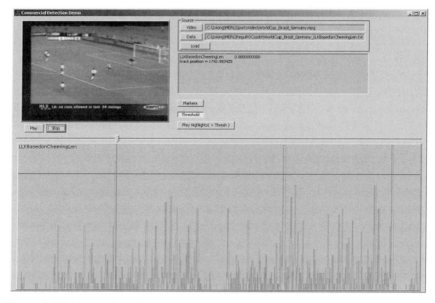

Figure 3.37   A snapshot of the interface of our system. The horizontal line imposed on the curve is the threshold.

3.5   Conclusions

This chapter has presented a sports highlights extraction framework that is built on a hierarchical representation that includes play/break segmentation, audio-visual marker detection, audio-visual marker association, and finer-resolution highlight classification. It decomposes the semantic and subjective concepts of "sports highlights" to events at different layers. As we go down the representation hierarchy, the events are less subjective and easier to model. The key component in this framework is the detection of audio and visual objects that serve as the bridge between the observed video signal and the semantics. It is a deviation from the "feature extraction + classification" paradigm for multimedia modeling, especially when the features are global features such as color histograms. Note that visual object detection also uses image features, but these features represent localized features (e.g., the boundary between the goalpost and the background) and spatial configuration of these local features (e.g., two vertical bars of the goalpost are at opposite sides of the pattern and have a horizontal bar in between). The experimental results have confirmed the advantage of this approach.

Chapter 4 | Video Structure Discovery Using Unsupervised Learning

4.1 Motivation and Related Work

We proposed a hierarchical representation for unscripted content as shown in Figure 4.1 [51]. It is based on the detection of domain-specific key audio-visual objects (audio-visual markers) that are indicative of the "highlight" or "interesting" events. The detected events can also be ranked according to a chosen measure, which would allow a generation of summaries of desired length [26]. In this representation, for each domain the audio-visual markers are chosen manually based on intuition.

For scripted content, the representation framework is based on the detection of the semantic units. Past work has shown that the representation units starting from the "key frames" up to the "groups" can be detected using unsupervised analysis. However, the highest-level representation unit requires content-specific rules to bridge the gap between semantics and the low-/mid-level analysis.

For unscripted content, the representation framework is based on the detection of specific events. Past work has shown that the play/break representation for sports can be achieved by an unsupervised analysis that brings out repetitive temporal patterns [6]. However, the rest of the representation units require the use of domain knowledge in the form of supervised audio-visual object detectors that are correlated with the events of interest. This necessitates a separate analysis framework for each domain in which the key audio-visual objects are chosen based on intuition. However, what is more desirable is a content-adaptive analysis and representation framework that postpones content-specific processing to as late a stage as possible. A content-adaptive solution is easier to realize in hardware for consumer electronic devices because the same hardware and software can handle multiple genres. Realizing such a framework poses the following challenges:

- Can we come up with a representation framework for unscripted content that postpones the use of the domain knowledge to a stage as late as possible?
- What kind of patterns would need to be discovered to support such a representation framework?
- Can such a framework help in the systematic choice of the key audio-visual objects for events of interest?

4.2 Proposed Inlier/Outlier–Based Representation for "Unscripted" Multimedia Using Audio Analysis

With the knowledge of the domain of the unscripted content, one can come up with an analysis framework with supervised learning tools for the generation of the hierarchical representation of events in unscripted content for summarization as shown in Figure 4.1.

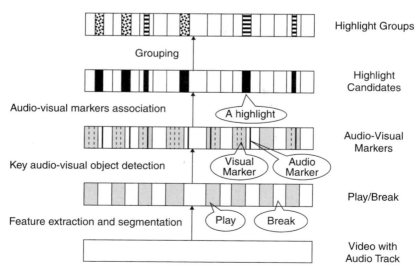

Figure 4.1 A hierarchical video representation for unscripted content.

In this representation framework, except for the analysis for play/break segmentation, the rest of the analysis requires content-specific processing since it involves specific supervised learning models for audio-visual markers detection. Instead, what is desirable is a framework that postpones such content-specific processing to as late a stage as possible. Postponing content-specific analysis to a later stage makes hardware realizations of the early common analysis easier than having specific realizations for different genres. Also, such an analysis framework helps in acquiring domain knowledge without characterizing the domain beforehand with intuition or partial domain knowledge. Hence, we propose a content adaptive analysis and representation framework that does not require any a priori knowledge of domain of the unscripted content. It is aimed toward an inlier/outlier–based representation of the content for event discovery and summarization as shown in Figure 4.2. It is motivated by the observation that "interesting" events in unscripted multimedia occur sparsely in a background of usual or "uninteresting" events. Some examples of such events are as follows:

- **Sports.** A burst of overwhelming audience reaction in the vicinity of a highlight event in a background of commentator's speech

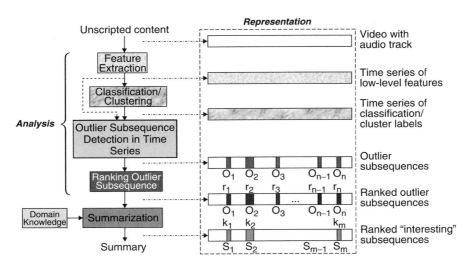

Figure 4.2 Proposed event discovery framework for analysis and representation of unscripted content for summarization.

- *Surveillance*. A burst of motion and screaming in the vicinity of a suspicious event in a silent or static background

Based on the aforementioned observations, we focus on audio analysis for the inlier/outlier–based temporal segmentation. Some of the other reasons for choosing audio analysis include the following:

- Detection of audio events from low-level features takes us closer to the semantics than detection of video events would. For instance, one can measure "interestingness" by simply measuring the audience reaction to the event irrespective of the event itself. Similar video analysis to understand events is difficult to achieve.
- To achieve the same level in semantic hierarchy, audio analysis is computationally more efficient than is video analysis.
- Since we measure the reaction to an event using audio analysis rather than the event itself, one can imagine a unified analysis framework for handling different genres. For instance, audience reaction is common to many sports and is a good measure of "interestingness," even though the meaning of "interesting" changes from sport to sport. Video analysis, on the other hand, tends to be content-specific as it measures the event directly and a completely unified framework is not possible. For instance, a key video object such as the catcher pose that marks the beginning of every pitch is specific to baseball content.

We briefly describe the role of each component in the proposed content-adaptive analysis framework as follows.

- *Feature extraction*. In this step, low-level features are extracted from the input content in order to generate a time series from which events are to be discovered. For example, the extracted features from the audio stream, could be Mel frequency cepstral coefficients (MFCC) or any other spectral/cepstral representation of the input audio.
- *Classification/Clustering*. In this step, the low-level features are classified using supervised models for classes that span the whole domain to generate a discrete time series of midlevel classification/clustering labels. One could also discover events from this sequence of discrete labels. For example, Gaussian mixture models (GMMs) can be used to classify every frame of audio into one of the following five audio classes, which span most of the sounds in sports audio: applause, cheering, music, speech, and speech with music. At this level, the input unscripted content is represented by a time series

of midlevel classification/cluster labels. These semantic labels can be thought of as coarser representations of the corresponding low-level features both in frequency and in time.

- *Detection of subsequences that are outliers in a time series*. In this step, we detect outlier subsequences from the time series of low-level features or midlevel classification labels motivated by the observation that "interesting" events are unusual events in a background of "uninteresting" events. At this level, the input content is represented by a temporal segmentation of the time series into inlier and outlier subsequences. The detected outlier subsequences are illustrated in Figure 4.2 as O_i, $1 \leq i \leq n$.

- *Ranking outlier subsequences*. To generate summaries of desired length, we rank the detected outliers with respect to a measure of statistical deviation from the inliers. At this level, the input content is represented by a temporal segmentation of the time series into inlier and ranked outlier subsequences. The ranks of detected outlier subsequences are illustrated in Figure 4.2 as r_i, $1 \leq i \leq n$.

- *Summarization*. The detected outlier subsequences are statistically unusual. All unusual events need not be interesting to the end user. Therefore, with the help of domain knowledge, we prune the outliers to keep only the interesting ones and modify their ranks. For example, commercials and highlight events are both unusual events and hence using domain knowledge in the form of a supervised model for audience reaction sound class will help in getting rid of commercials from the summary. At this level, the input content is represented by a temporal segmentation of the time series into inlier and ranked "interesting" outlier subsequences. The "interesting" outlier subsequences are illustrated in Figure 4.2 as S_i, $1 \leq i \leq m$ with ranks k_i. The set of "interesting" subsequences (S_i's) is a subset of outlier subsequences (O_i's).

In the following section, we will describe in detail the first two analysis blocks: feature extraction and the audio classification framework.

4.3 Feature Extraction and the Audio Classification Framework

The feature extraction block and the audio classification block help represent the input audio stream from the unscripted content using two different time series representations. The feature extraction block extracts a frequency domain representation for the input audio, resulting in a time series of continuous feature vectors. The audio classification block classifies each continuous feature vector into one of several audio classes, resulting in a time series of discrete audio classification labels. In this section, we describe each of these blocks in more detail.

4.3.1   FEATURE EXTRACTION

The extraction of the following two types of audio features was used:

- Mel frequency cepstral coefficients (MFCC)
- Modified discrete cosine transform (MDCT) coefficients

4.3.2   MEL FREQUENCY CEPSTRAL COEFFICIENTS (MFCC)

The human ear resolves frequencies nonlinearly across the audio spectrum and empirical evidence from speech recognition applications show that systems that work with features derived in a similar manner improve recognition performance. Since structure discovery from multimedia also relies on recognition of some semantic concepts from low-level features, MFCC features are a natural choice for recognition tasks based on audio features.

To extract MFCC features, input audio is divided into overlapping frames of duration 30 ms with 10 ms overlapping for consecutive frames. Each frame is then multiplied by a hamming window function:

$$w_i = 0.5 - 0.46\left(\cos\left(\frac{2\pi}{N}\right)\right), \ 1 \le i \le N \qquad (4.3.1)$$

where N is the number of samples in the window. After performing FFT on each windowed frame, Mel frequency cepstral coefficients (MFCC) are calculated using the following discrete cosine transform:

$$C_n = \sqrt{\frac{2}{K}} \sum_{i=1}^{K} \log S_i \times \cos\left(n\left(i - \frac{1}{2}\right)\pi/K\right), \ n = 1, 2, \ldots, L \quad (4.3.2)$$

where K is the number of subbands, and L is the desired length of cepstrum. Usually L is chosen as 10 for dimensionality reduction. S_i's, $1 \le i \le K$, represent the filter bank energy after passing through the triangular band-pass filters. The band edges for these band-pass filters are corresponding to

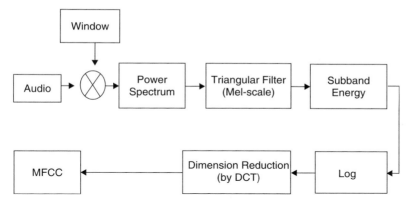

Figure 4.3 MFCC feature extraction.

Mel frequency scale (linear scale below 1 kHz and logarithmic scale above 1 kHz). Figure 4.3 summarizes the MFCC extraction process.

This feature extraction process takes care of energy normalization in the following way. Let us assume two audio clips A_1 and A_2 ($A_2 = m \times A_1$). Then, the subband energy values (S_i's) for clip A_2 are correspondingly scaled by a factor of m. Taking the discrete cosine transform (DCT) on the logarithm of the subband energy values packs the scaling factor into the DC component after the transform. Then, by not including the DC component into the feature vector, we would have achieved energy normalization.

Also, note that dimensionality reduction in MFCC feature extraction is achieved by projecting the subband energies onto the cosine basis. Since the DCT is an approximation to Karhunen Loeve transform (KLT), it packs the energy into few coefficients, and by discarding higher-order coefficients with small energy we reduce dimensionality while preserving most of the energy.

In the following subsection, we describe the modified discrete cosine transform feature extraction procedure.

4.3.3 MODIFIED DISCRETE COSINE TRANSFORM (MDCT) FEATURES FROM AC-3 STREAM

The modified discrete cosine transform (MDCT) is another frequency domain representation of an audio signal and is used for transform coding of audio in AC-3 dolby codecs and MPEG-2 AAC codecs. If a reduced dimensionality representation of these transform coefficients can be used for recognition tasks, it would yield some savings in computation over

MFCC features, which are extracted from the time domain representation of an audio signal. Also, the triangular band-pass filters in MFCC extraction may not suit general audio classification since some spectrogram structure is lost when energy is summed up within each band. Therefore, we are motivated to use the MDCT coefficients extracted from the compressed domain for the classification task.

The MDCT is defined as follows:

$$X(m) = \sum_{k=0}^{n-1} f(k)x(k) \cos\left(\frac{\pi}{2n}\left(2k+1+\frac{n}{2}\right)(2m+1)\right), m = 0, 1, \ldots \frac{n}{2} - 1$$

(4.3.3)

These coefficients are available in the AC-3 bit stream, and one can simply parse the bit stream to extract them. Figure 4.4 shows the AC-3 frame structure at 48-KHz sampling rate. During encoding, a block of 512 time domain samples (corresponding to 5.3 ms) is transformed to generate 256 frequency domain MDCT coefficients. A frame of AC-3 has six such MDCT blocks. We average the absolute values of these MDCT blocks to get one feature vector of dimension 256 for every 31.8 ms. Note that we take the absolute value of MDCT coefficients in each block before computing the mean. This is essential to preserve the underlying frequency structure even after taking the mean. To understand this, let us look at the MDCT coefficients for a synthetic audio signal, $A(t)$, with two tones given as follows:

$$A(t) = \cos(2\pi 5000t) + \sin(2\pi 8000t)$$

(4.3.4)

Figure 4.5 shows the MDCT coefficients of the first block. There are peaks at the 53rd coefficient (corresponding to 5 kHz) and the 85th

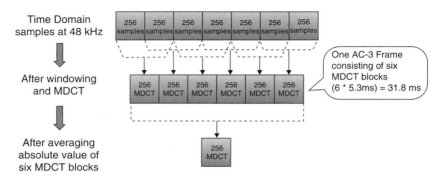

Figure 4.4   MFCC feature extraction.

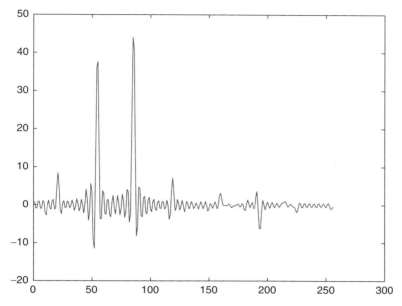

Figure 4.5 MDCT coefficients for the first block.

coefficient (corresponding to 8 kHz) as expected. Figure 4.6 shows the MDCT coefficients of the second block. Again, there are peaks at the same coefficients, but the signs are reversed. This is because MDCT coefficients contain the phase information as well. Therefore, computing the mean, without taking this fact into account, will destroy the underlying frequency structure as shown in Figure 4.7. However, when we take the absolute values of MDCT coefficients for computing the mean, the frequency structure is preserved as shown in Figure 4.8.

The MDCT coefficients for each frame extracted in this manner have to be normalized with respect to energy as in the case of MFCC extraction. As mentioned earlier, in MFCC, energy normalization is achieved by removing the DC coefficient of the DCT performed after log operation. Each frame's MDCT is normalized by the average energy of all the frames that correspond to 1 second in the vicinity of that frame. The energy normalization procedure is shown in Figure 4.9.

After energy normalization, dimensionality reduction is achieved by projecting each frame's MDCT onto N most significant singular value decomposition (SVD) basis vectors. The value of N can be determined by computing the effective rank of the MDCT coefficients from all the

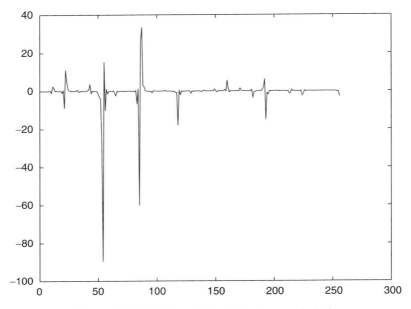

Figure 4.6   MDCT coefficients for the second block.

clips that were used for computing the SVD basis. Let X be the $n \times p$ data matrix with n being equal to the total number of MDCT frames from which the basis needs to be computed. p is the dimensionality of each MDCT coefficient vector and is equal to 256 in this case. Using singular value decomposition of X as

$$X = U \Sigma V^T \qquad (4.3.5)$$

where U is a $n \times n$ matrix (called the left singular matrix) with each column representing a left singular vector. Σ is an $n \times p$ diagonal matrix with singular values $(\sigma_1, \sigma_2, \ldots, \sigma_p)$ along the main diagonal. V is a $p \times p$ matrix (called the right singular matrix) with each column representing a right singular vector. One can compute a rank q approximation, X_q for the matrix X using the following equation:

$$\mathbf{X_q} = \sum_{k=1}^{q} \sigma_k \mathbf{U_k} \mathbf{V_k}^T \qquad (4.3.6)$$

If most of the energy in the data resides in a subspace of dimension $r \leq min(n, p)$, then the corresponding approximation $\mathbf{X_r}$ would be

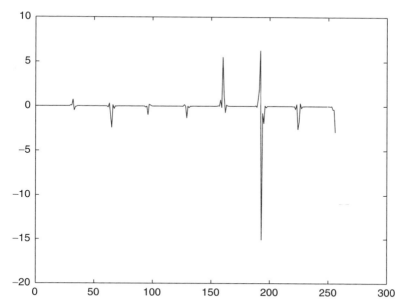

Figure 4.7 MDCT coefficients for a frame from the mean of six MDCT blocks without absolute value operation.

good enough. Then, we could use the first r singular vectors to project the data onto that subspace, thereby reducing the dimensionality of the data.

Figure 4.10 shows the approximation error as one varies the subspace dimensions. The Y-axis plots $\frac{||X-X_q||}{||X||}$ (η) for every candidate subspace dimension. Here $||X||$ represents the norm of the matrix and is equal to the sum of squares of all the terms. For instance, for an approximation error of 0.1, we can choose the effective rank to be 21. This means that 90% of the energy in the input 256 dimensional vector can be packed into the first 21 dimensions. One can also choose the effective rank by looking at the energy captured in the rank q approximation (X_q) relative to the original energy in X ($\mu = \frac{||X_q||}{||X||}$). Figure 4.11 shows this quantity as a function of q. This low dimensional representation is useful in two ways:

- It is computationally efficient to work with a smaller number of dimensions.
- For the same training data size, the probability density function (PDF) learned from a smaller dimensional representation of the data is more reliable than a PDF learned from a larger dimensional representation of the data.

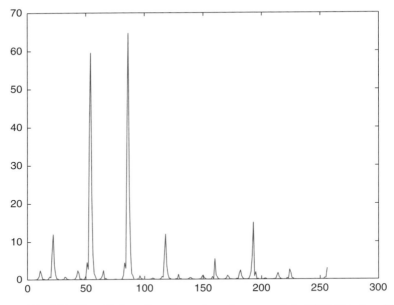

Figure 4.8   MDCT coefficients for a frame from the mean of six MDCT blocks with the absolute value operation.

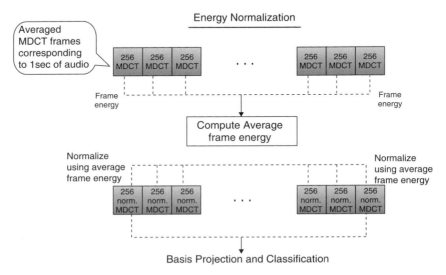

Figure 4.9   Energy normalization for MDCT coefficients.

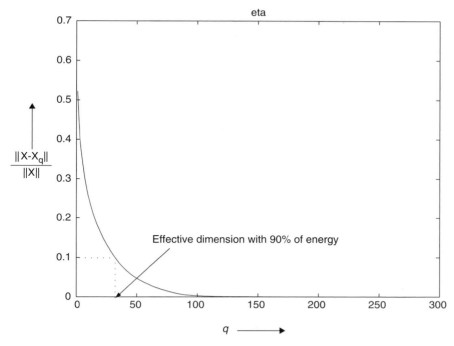

Figure 4.10 η Vs q for effective rank.

Now that we have described two different feature extraction procedures for the input audio, we will describe the audio classification framework based on these features.

4.3.4   AUDIO CLASSIFICATION FRAMEWORK

The audio classification framework takes as input a continuous time series of low-level audio features (MFCC/MDCT) and outputs a discrete time series of midlevel semantic labels. The role of the audio classification framework is to allow for a coarser representation of the low-level features time series both in frequency and in time. For instance, in sports audio we have chosen five audio classes based on intuition, and all the low-level feature vectors are classified as belonging to one of the following five audio classes: applause, cheering, music, speech, and speech with music. The first two classes characterize the audience reaction sounds, and the last three cover rest of the sounds in the sports domain [26].

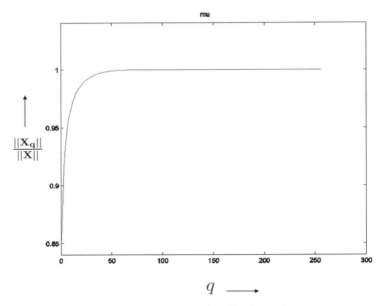

Figure 4.11 μ Vs q for effective rank.

Figure 4.12 illustrates the audio classification framework based on the extracted features. We collected several hours of training data for each of the chosen audio classes from broadcast content. Then, low-level features were extracted from each of the training clips as described in the previous section. The distribution of low-level features was parameterized by a Gaussian mixture model (GMM). A GMM is a generative probabilistic model that can approximate any PDF with sufficient training data. Let Y be an M-dimensional random vector to be modeled using a Gaussian mixture distribution. Let K denote the number of Gaussian mixtures, and we use the notation π, μ, and R to denote the parameter sets $\{\pi_k\}_{k=1}^{K}$, $\{\mu_k\}_{k=1}^{K}$, and $\{R_k\}_{k=1}^{K}$, respectively, for mixture coefficients, means, and variances. The complete set of parameters are then given by K and $\theta = (\pi, \mu, R)$. The log of the probability of the entire sequence $Y = \{Y_n\}_{n=1}^{N}$ is then given by

$$\log p_y(y|K, \theta) = \sum_{n=1}^{N} \log \left(\sum_{k=1}^{K} p_{y_n|x_n}(y_n|k, \theta)\pi_k \right) \qquad (4.3.7)$$

The objective is then to estimate the parameters K and θ. We use Rissanen's minimum description length principle and expectation maximization algorithm to learn these parameters from the training data

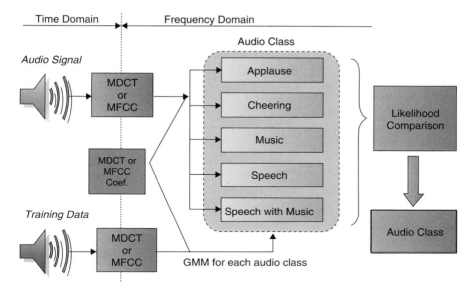

Figure 4.12 Audio classification framework.

for each class. The method trades off between model complexity and likelihood of the data under the model [26].

Once model parameters for each audio class are learned during training, the likelihood of an input low-level feature vector (to be classified during testing) under all the trained models is computed. It is classified as belonging to one of the sound classes based on a maximum likelihood criterion.

In the following section, we describe the proposed time series analysis framework, which uses the output of feature extraction and classification blocks.

4.4 Proposed Time Series Analysis Framework

We treat the sequence of low-/midlevel features extracted from the audio as a time series and identify subsequences that are outliers. This kind of inlier/outlier–based temporal segmentation of the input time series is motivated by the observation that "interesting" events in unscripted multimedia occur sparsely in a background of usual or "uninteresting" events.

Outlier subsequence detection is at the heart of the proposed framework. Examples of such events are as follows:

- **Sports**. A burst of overwhelming audience reaction in the vicinity of a highlight event in a background of commentator's speech
- **Surveillance**. A burst of motion and screaming in the vicinity of a suspicious event in a silent or static background

This motivates us to formulate the problem of discovering "interesting" events in multimedia as that of detecting outlier subsequences or "unusual" events by statistical modeling of a stationary background process in terms of low-/midlevel audio-visual features. Note that the background process may be stationary only for a short period of time and can change over time. This implies that background modeling has to be performed adaptively throughout the content. It also implies that it may be sufficient to deal with one background process at a time and detect outliers. In the following subsection, we elaborate on this more formally.

4.4.1   PROBLEM FORMULATION

Let p_1 represent a realization of the "usual" class ($\mathbf{P_1}$), which can be thought of as the background process. Let p_2 represent a realization of the "unusual" class $\mathbf{P_2}$, which can be thought of as the foreground process. Given any time sequence of observations or low-level audio-visual features from the two the classes of events ($\mathbf{P_1}$ and $\mathbf{P_2}$), such as

$$\cdots p_1 p_1 p_1 p_1 p_1 p_2 p_2 p_1 p_1 p_1 \cdots$$

then the problem of outlier subsequence detection is that of finding the times of occurrences of realizations of $\mathbf{P_2}$ without any a priori knowledge about $\mathbf{P_1}$ or $\mathbf{P_2}$.

To begin with, the statistics of the class $\mathbf{P_1}$ are assumed to be stationary. However, there is no assumption about the class $\mathbf{P_2}$. The class $\mathbf{P_2}$ can even be a collection of a diverse set of random processes. The only requirement is that the number of occurrences of $\mathbf{P_2}$ is relatively rare compared to the number of occurrences of the dominant class. Note that this formulation is a special case of a more general problem—namely, clustering of a time series in which a single highly dominant process does not necessarily exist. We treat the sequence of low-/midlevel audio-visual features extracted from the video as a time series and perform a temporal segmentation to detect transition points and outliers from a sequence of observations.

4.4.2 *KERNEL/AFFINITY MATRIX COMPUTATION*

To detect outlier subsequences from an input time series of observations from $\mathbf{P_1}$ and $\mathbf{P_2}$, we first need to sample the input time series into subsequences. Then, we need to represent the sampled subsequences in a feature space for subsequent clustering and outlier detection. Since we are modeling the input time series using a generative model for $\mathbf{P_1}$ and $\mathbf{P_2}$, a natural way to represent the subsequences in a feature space is to use a sequence kernel based on an estimate of the underlying process generating the subsequences. From every subsequence, we estimate a statistical model for the underlying process generating the observations. Let us represent two such statistical models, λ_1 and λ_2, estimated from subsequences O_1 and O_2, respectively. The sequence kernel to compare these two subsequences is based on the distance metric defined as follows:

$$D(\lambda_1, \lambda_2) = \frac{1}{W_L}(\log P(O_1|\lambda_1) + \log P(O_2|\lambda_2)$$

$$- \log P(O_1|\lambda_2) - \log P(O_2|\lambda_1)) \qquad (4.4.1)$$

where W_L is the number of observations within a sampled subsequence. Note that this distance metric not only measures how well λ_1 explains the training data (O_1) from which it was estimated, but it also measures how well it explains the data in the other subsequence (O_2). If both the subsequences were generated from the process $\mathbf{P_1}$, then the models estimated from each of them would be able to explain the observations in the other equally well. This implies that the first two terms in the distance metric just defined would be close to the last two terms, causing the distance to be as small as desired. On the other hand, if one of the subsequences has a burst of observations from the process $\mathbf{P_2}$, then the corresponding model would not be able to explain the data in the other subsequence that has observations only from $\mathbf{P_1}$, causing the distance metric value to be larger. Using this distance metric, we can compute a distance matrix, D, using pair-wise distances between every pair of the sampled subsequences. Then, we can compute the affinity/kernel matrix (A) for the whole time series using the distance matrix entries $(d(i, j))$'s. Each element, $A(i, j)$, in the affinity matrix is $e^{\frac{-d(i, j)}{2\sigma^2}}$ where σ is a parameter that controls how quickly affinity falls off as distance increases. By computing the affinity matrix for the time series in this manner, we have essentially embedded the subsequences in a high dimensional space using the sequence kernel. The configuration of the points in that high dimensional space is captured by the pair-wise similarity between the subsequences in that space.

Figure 4.13(a) shows a time series of discrete symbols (1, 2, 3) from two processes. Observations from the dominant process, $\mathbf{P_1}$, and observations from the other process, $\mathbf{P_2}$, are shown. Figure 4.14 shows the zoomed-in view of the same time series. Note that the input time series is a sequence of symbols from the alphabet 1, 2, 3. Figure 4.13(b) shows the affinity matrix computed from subsequences sampled from this time series. The sequence kernel used in this case was based on a hidden Markov model (HMM). The dark regions in the affinity matrix correspond to times of occurrences of observations from the process $\mathbf{P_2}$ and imply that models estimated using subsequences in those parts of the time series were very different from the others.

To detect such "outlier" subsequences using the affinity matrix, we resort to the graph theoretical approach to clustering. Before we develop the complete framework for detecting outlier subsequences, we review the related theoretical background on the graph theoretical approach to clustering.

4.4.3 SEGMENTATION USING EIGENVECTOR ANALYSIS OF AFFINITY MATRICES

Segmentation using eigenvector analysis has been proposed for images in [60]. This approach to segmentation is related to the graph theoretic formulation of grouping. The set of points in an arbitrary feature space is represented as a weighted undirected graph where the nodes of the graph are points in the feature space and an edge is formed between every pair of nodes. The weight on each edge is the similarity between the two nodes connected by that edge. Let us denote the similarity between nodes i and j as $w(i, j)$.

To understand the partitioning criterion for the graph, let us consider partitioning it into two groups: $\mathbf{A}$ and $\mathbf{B}$ and $A \cup B = V$:

$$N_{cut}(A, B) = \frac{cut(A, B)}{asso(A, V)} + \frac{cut(A, B)}{asso(B, V)} \qquad (4.4.2)$$

where

$$cut(A, B) = \sum_{i \in A, j \in B} w(i, j) \qquad (4.4.3)$$

$$asso(A, V) = \sum_{i \in A, j \in V} w(i, j) \qquad (4.4.4)$$

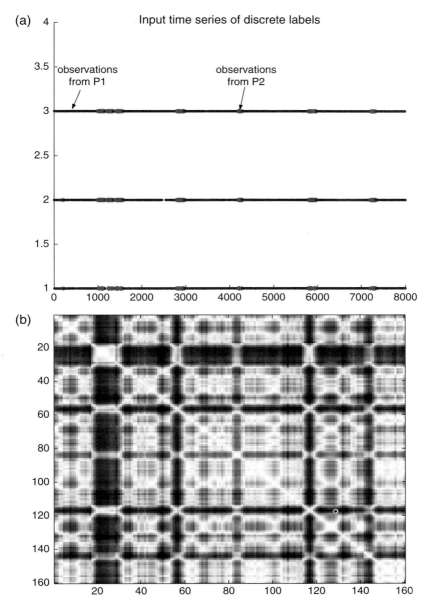

Figure 4.13   Input time series. (a) Time along X-axis and symbols (1, 2, 3) along Y-axis.
(b) The computed affinity matrix.

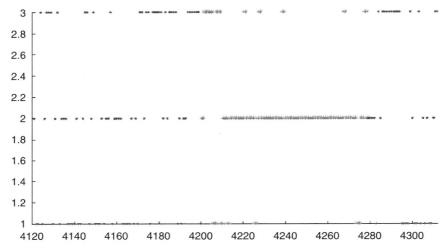

Figure 4.14   Zoomed-in view of the input time series from $\mathbf{P}_1$ and $\mathbf{P}_2$, time along X-axis, and symbols (1, 2, 3) along Y-axis.

It has been shown that minimizing N_{cut} minimizes similarity between groups while maximizing association within individual groups [60]. Shi and Malik [60] show that

$$min_x N_{cut}(x) = min_y \frac{y^T (D - W)y}{y^T D y} \qquad (4.4.5)$$

with the condition $y_i \epsilon \{-1, b\}$. Here, W is a symmetric affinity matrix of size $N \times N$ (consisting of the similarity between nodes i and j, $w(i, j)$ as entries) and D is a diagonal matrix with $d(i, i) = \sum_j w(i, j)$. x and y are cluster indicator vectors—that is, if $\mathbf{y(i)}$ equals -1, then feature point 'i' belongs to cluster A, otherwise it belongs to cluster B. It has also been shown that the solution to the preceding equation is the same as the solution to the following generalized eigenvalue system if y is relaxed to take on real values:

$$(D - W)y = \lambda D y \qquad (4.4.6)$$

This generalized eigenvalue system is solved by first transforming it into the standard eigenvalue system by substituting $z = D^{\frac{1}{2}} y$ to get

$$D^{-\frac{1}{2}} (D - W) D^{-\frac{1}{2}} z = \lambda z \qquad (4.4.7)$$

It can be verified that $z_0 = D^{\frac{1}{2}} \vec{1}$ is a trivial solution with eigenvalue equal to 0. The second generalized eigenvector (the smallest nontrivial solution) of this eigenvalue system provides the segmentation that optimizes N_{cut} for two clusters. We will use the term "the cluster indicator vector" interchangeably with "the second generalized eigenvector of the affinity matrix."

4.4.4 PAST WORK ON DETECTING "SURPRISING" PATTERNS FROM TIME SERIES

The problem of detecting "surprising" patterns from time series has been looked at by many researchers from a variety of domains, including bioinformatics, computer and network security, finance, and meteorology. Each of these domains has its own specific data stream, and even though the notion of an outlier is quite intuitive, its definition seems to change depending on the domain, as we shall see.

Jagadish et al. [52] define the notion of a deviant in a time series based on information theory. Specifically, if the removal of a point P results in a sequence that can be represented by histogram more efficiently than the original sequence, then the point P is considered to be a deviant. The authors propose an efficient algorithm to find the subset of deviants from the whole time series and show its effectiveness on one-dimensional time series data from a census database and a financial time series. The proposed algorithm works on univariate time series and hence cannot be applied to the time series of low-level multivariate cepstral/spectral features that we have at our hand.

Muthukrishnan et al. [53] propose an algorithm to find deviants from data streams. Unlike the techniques proposed by Jagadish et al. [52], which only work on stored time series data, this work finds deviants from streaming data. The proposed scheme has been shown to be effective in detecting deviants from simple network management protocol (SNMP) network traffic data. Both of these proposed algorithms would not be useful to the problem at hand as we are interested in detecting a burst of multivariate observations that are generated from a different process than the usual and not in the detection of a single deviant from a waveform.

Dasgupta et al. [54] propose an immunology-based approach to novelty detection in time series. It is inspired by the negative selection mechanism of the human immune system, which discriminates between *self* and *other*. From a quantized representation of the reference normal univariate time series data, a multiset of strings, S, is extracted to represent the *self*. Then, a set of detectors is generated that does not match any of the string in S. So given a new data set, if any of the extracted strings match one of the detectors, a novelty is declared. The proposed algorithm cannot be applied directly to the problem at hand as it only works with univariate time series, and a reference data set is required to learn the strings for the *self*. Also, the strings for *self* are only learned from discrete univariate time series.

Ma et al. [55] propose an online novelty detection algorithm using support vector regression to model the current knowledge of the usual. The proposed system requires some training data to perform support vector regression. Then, based on the learned model, a confidence is output on the detected novel events. Since this scheme requires careful selection of training data for modeling the usual using support vector regression, it cannot be applied to our problem.

Shahabi et al. [56] propose a method based on wavelet decomposition of time series. The proposed system detects dramatic signal shifts at various scales in the high-pass bands. The system detects trends at various scales in the low-pass bands. Since the definition of surprise is limited to sudden signal changes, it is not useful to detect a burst of observations that are different from the usual.

Chakrabarti et al. [57] propose a scheme to discover surprising patterns from Boolean market basket data. The notion of surprise is based on the number of bits needed to encode an item set sequence on using a specific coding scheme. The encoding scheme takes relatively fewer bits to encode item sets that have steady correlation between items while taking more bits to code a possibly surprising correlation. Since the proposed scheme is for Boolean market basket data, it doesn't apply for the type of time series data we will be dealing with.

Brutlag [58] proposes a scheme to detect aberrant behavior for network service monitoring. The algorithm predicts the value of time series one step into the future based on a number of past samples. Then, an aberration is declared when an observed value is too deviant from the predicted value. Again, this scheme detects surprising patterns from univariate time series data and cannot be directly applied to outlier subsequence detection from multivariate time series.

Keogh et al. [59] propose a method called TARZAN for detecting novelties from time series. It is based on converting the input reference univariate time series into a symbolic string, and a suffix tree is computed to count the number of occurrences of each substring in the sequence of symbols. Then, given a new time series a pattern is declared surprising if the frequency of its occurrence differs substantially from that estimated from the suffix tree of the reference time series. This scheme is also not suitable for our problem as it works with univariate time series and requires a reference time series. Furthermore, the procedure for discretizing and symbolizing real values in time series can cause the loss of meaningful patterns in the original time series.

In summary, past work on detecting surprising patterns from time series has mostly concentrated on working with univariate time series. However, we are mainly interested in detecting outlier subsequences from a multivariate time series of low-level cepstral/spectral features. This is because semantic concepts in audio are best captured by multivariate features from the spectral/cepstral domain and not in the time domain. Furthermore, past work has concentrated on detecting outliers as points on the time series whose values deviate from the values in a reference time series. However, we are interested in detecting outliers that are segments of a time series whose statistics are very different from a usual dominant background process. For example, the statistic could be the probability distribution function (PDF) estimate of the underlying process generating the observations. Also, unlike past work on time series, we don't assume any reference time series beforehand. Therefore, applying techniques from time series literature directly to time domain waveform would not help us detect a burst of audience reaction, for example.

Hence, we propose the following outlier subsequence detection framework, which works with a time series of multivariate feature vectors. It is based on eigenvector analysis of the affinity matrix constructed from statistical models estimated from the subsequences of the time series of multivariate features.

4.4.5 PROPOSED OUTLIER SUBSEQUENCE DETECTION IN TIME SERIES

Given the problem of detecting times of occurrences of P_1 and P_2 from a time series of observations from P_1 and P_2, we propose the following time series clustering framework:

(1) Sample the input time series on a uniform grid. *Let each time series sample at index "i" (consisting of a sequence of observations) be referred to as a context, C_i.* This definition of a context is inspired from lossless image compression literature where a context is defined as a causal neighborhood of the current pixel. However, here a context refers to both a causal and a noncausal neighborhood in time.

(2) Compute a statistical model M_i from the time series observations within each C_i.

(3) Compute the affinity matrix for the whole time series using the context models and a commutative distance metric $(d(i, j))$ defined between two context models $(M_i$ and $M_j)$. Each element, $A(i, j)$, in the affinity matrix is $e^{\frac{-d(i,j)}{2\sigma^2}}$, where σ is a parameter that controls how quickly affinity falls off as distance increases.

(4) The computed affinity matrix represents an undirected graph where each node is a context model and each edge is weighted by the similarity between the nodes connected by it. Then, we can use a combination of the normalized cut solution and the modified normalized cut solution to identify distinct clusters of context models and "outliers context models" that do not belong to any of the clusters. Note that the second generalized eigenvector of the computed affinity matrix is an approximation to the cluster indicator vector for bipartitioning the input graph, as discussed in Section 4.4.3.

Figure 4.15 illustrates the proposed framework. The portion of the figure (b) is a detailed illustration of the two blocks: (clustering and outlier

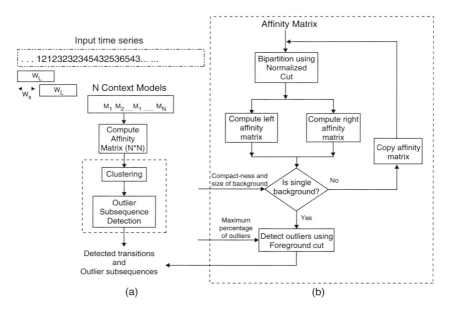

Figure 4.15   Proposed outlier subsequence detection framework.

detection) in figure (a). In this framework, there are two key issues: the choice of the type of statistical model for the context and the choice of the two parameters, W_L and W_S. The choice of the statistical model for the time series sample in a context would depend on the underlying background process. A simple unconditional PDF estimate would suffice for a memoryless background process. However, if the process has some memory, the chosen model would have to account for it. For instance, a hidden Markov model would provide a first-order approximation.

The choice of the two parameters (W_L and W_S) would be determined by the confidence with which a subsequence is declared to be an outlier. The size of the window W_L determines the reliability of the statistical model of a context.The size of the sliding factor, W_S, determines the resolution at which the outlier is detected.

Before we discuss how these parameters affect the performance, we will develop the theory for clustering subsequences of time series by working with synthetic time series data. In the next section, we describe our generative model for synthetic time series.

4.4.6   *GENERATIVE MODEL FOR SYNTHETIC TIME SERIES*

To test the effectiveness of the proposed outlier subsequence detection framework, we generate a synthetic time series using the generative model shown in Figure 4.16.

In this framework, we have a generative model for both $\mathbf{P_1}$ and $\mathbf{P_2}$, and the dominance of one over the other can also be governed by a probability parameter ($P(p_2)$). With probability $P(p_2)$, process $\mathbf{P_2}$ generates a burst of observations. The length of the burst is a uniform random variable and determines the percentage of observations from $\mathbf{P_2}$ in a given context.

There are four possible scenarios one can consider with the proposed generative model for label sequences:

- *Case 1*. Sequence completely generated from $\mathbf{P_1}$. This case is trivial.
- *Case 2*. Sequence dominated by observations from one background process ($\mathbf{P_1}$) with only one foreground process ($\mathbf{P_2}$)—that is, $P(\mathbf{P_1}) >> P(\mathbf{P_2})$.
- *Case 3*. Sequence dominated by observations from one background process ($\mathbf{P_1}$) and many foreground processes ($\mathbf{P_2}, \mathbf{P_3}, \mathbf{P_4}$)—that is, $P(\mathbf{P_1}) >> P(\mathbf{P_2}) \approx P(\mathbf{P_3}) \approx P(\mathbf{P_4})$.

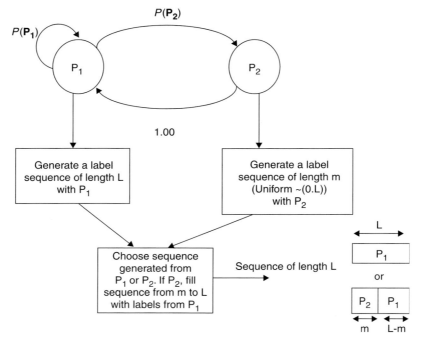

Figure 4.16   Generative model for synthetic time series with one background process and one foreground process.

- *Case 4*. with observations from P_1 and P_2 with no single dominant class with a number of foreground processes—that is, $P(P_1) \approx P(P_2)$ and $(P(P_1) + P(P_2)) >> P(P_3) + P(P_4)$.

Figure 4.17 shows the three cases of interest.

4.4.7   PERFORMANCE OF THE NORMALIZED CUT FOR CASE 2

In this section, we show the effectiveness of the normalized cut for case 2, when $P(P_1) >> P(P_2)$. Without loss of generality, let us consider an input discrete time series with an alphabet of three symbols $(1, 2, 3)$ generated from two HMMs (P_1 and P_2).

The parameters of P_1—(the state transition matrix (A), the state observation symbol probability matrix (B), and the initial state probability

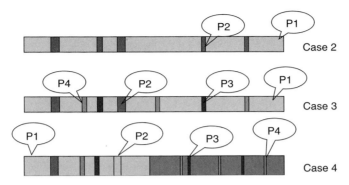

Figure 4.17 Cases of interest.

matrix (π))—are as follows:

$$\mathbf{A_{P_1}} = \begin{pmatrix} 0.3069 & 0.0353 & 0.6579 \\ 0.0266 & 0.9449 & 0.0285 \\ 0.5806 & 0.0620 & 0.3573 \end{pmatrix}$$

$$\mathbf{B_{P_1}} = \begin{pmatrix} 0.6563 & 0.2127 & 0.1310 \\ 0.0614 & 0.0670 & 0.8716 \\ 0.6291 & 0.2407 & 0.1302 \end{pmatrix}$$

$$\pi_{P_1} = \begin{pmatrix} 0.1 & 0.8 & 0.1 \end{pmatrix}$$

The parameters of $\mathbf{P_2}$ are

$$\mathbf{A_{P_2}} = \begin{pmatrix} 0.9533 & 0.0467 \\ 0.2030 & 0.7970 \end{pmatrix}$$

$$\mathbf{B_{P_2}} = \begin{pmatrix} 0.0300 & 0.8600 & 0.1100 \\ 0.3200 & 0.5500 & 0.1300 \end{pmatrix}$$

$$\pi_{P_2} = \begin{pmatrix} 0.8 & 0.2 \end{pmatrix}$$

Then, using the generative model shown in Figure 4.16 with $P(\mathbf{P_1}) = 0.8$ and $P(\mathbf{P_2}) = 0.2$, we generate a discrete time series of symbols as shown in Figure 4.18(a).

We sample this series uniformly using a window size of $W_L = 200$ and a step size of $W_S = 50$. We use the observations within every context to estimate an HMM with two states. Using the distance metric defined

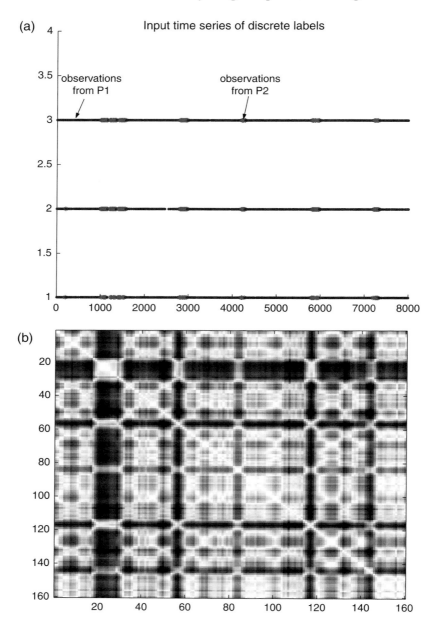

Figure 4.18   Performance of normalized cut on synthetic time series for case 2. (a) X-axis for time, Y-axis for symbol. (b) Affinity matrix. (c) X-axis for candidate normalized cut threshold, Y-axis for value of normalized cut objective function. (d) X-axis for time index of context model, Y-axis for cluster indicator value. (e) X-axis for time, Y-axis for symbol.

(c)

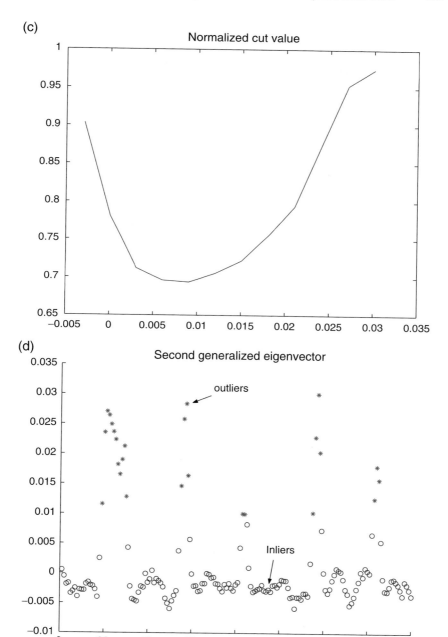

Figure 4.18 Cont'd

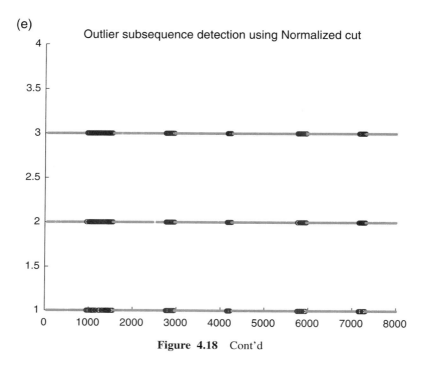

Figure 4.18   Cont'd

next for comparing two HMMs, we compute the distance matrix for the whole time series. Given two context models, λ_1 and λ_2 with observation sequences O_1 and O_2, respectively, we define

$$D(\lambda_1, \lambda_2) = \frac{1}{W_L} (\log P(O_1|\lambda_1) + \log P(O_2|\lambda_2)$$

$$- \log P(O_1|\lambda_2) - \log P(O_2|\lambda_1)) \qquad (4.4.8)$$

The computed distance matrix, D, is normalized to have values between 0 and 1. Then, using a value of $\sigma = 0.2$, we compute the affinity matrix, A, where $A(i, j) = e^{\frac{-d(i,j)}{2\sigma^2}}$. The affinity matrix is shown in Figure 4.18(b). We compute the second generalized eigenvector of this affinity matrix as a solution to the cluster indicator vector. Since the cluster indicator vector does not assume two distinct values, a threshold is applied on the eigenvector values to get the two clusters. To compute the optimal threshold, the normalized cut value is computed for the partition resulting from each candidate threshold between the range of eigenvector values. The optimal threshold is selected as the threshold at which the normalized cut value (the objective function) is at its minimum, as shown in Figure 4.18(c). The corresponding second generalized vector and its optimal partition are shown in Figure 4.18(d). The detected outliers are at times of occurrences of P_2. Figure 4.18(e) marks the detected outlier subsequences in the original time series based on the normalized cut. It can be observed that the outlier subsequences have been detected successfully without having to set any threshold manually. Also, note that since all outlier subsequences are from the same foreground process (P_2), the normalized cut solution was successful in finding the outlier subsequences. In general, as we shall see later, when the outliers are from more than one foreground process (case 3), the normalized cut solution may not perform as well. This is because each outlier can be different in its own way and it is not right to emphasize an association between the outlier cluster members as the normalized cut does.

In the following subsection, we show the performance of other competing clustering approaches for the same task of detecting outlier subsequences using the computed affinity matrix.

4.4.8 COMPARISON WITH OTHER CLUSTERING APPROACHES FOR CASE 2

As mentioned in Section 4.4.5 and shown in Figure 4.15, the clustering and outlier detection step follows construction of the affinity matrix (Step 3). Instead of using the normalized cut solution for clustering, one could also use one of the following three methods for clustering.

- Clustering using alphabet-constrained k-means
- Clustering based on a dendrogram
- Clustering based on factorization of the affinity matrix

Let us compare the performance of these competing clustering approaches for case 2.

4.4.8.1 Clustering Using in Alphabet-Constrained k-Means

Given the pair-wise distance matrix and the knowledge of the number of clusters, one can perform top-down clustering based on alphabet constrained k-means as follows. Since the clustering operation is performed in model space, the centroid model of a particular cluster of models is not merely the average of the parameters of cluster members. Therefore, the centroid model is constrained to be one of the models and is that model with minimum average distance to the cluster members.

Given that there is one dominant cluster and the distance matrix, we can use the following algorithm to detect outliers:

(1) Find the row in the distance matrix for which the average distance is at its minimum. This is the centroid model.

(2) Find the semi-Hausdorff distance between the centroid model and the cluster members. The semi-Hausdorff distance, in this case, is simply the maximum of all the distances computed between the centroid model and the cluster members. Hence, the semi-Hausdorff distance would be much larger than the average distance if there are any outliers in the cluster members.

(3) Remove the farthest model and repeat Step 2 until the difference between average distance and the Hausdorff distance is less than a chosen threshold.

(4) The remaining cluster members constitute the inlier models.

For more than one cluster, repeat Steps 1 to 3 on the complementary set, which does not include members of the detected cluster. Figure 4.19(a) shows the distance matrix values of the row that are corresponding to the centroid row. By using a threshold on the difference between the average distance and the Hausdorff distance, we detect outlier subsequences as shown in Figure 4.19(b).

4.4.8.2   Clustering Based on Dendrogram

Given the pair-wise distance matrix, one can perform bottom-up agglomerative clustering. At the start, each member is considered to be a distinct cluster. By merging the two closest clusters at every level until there is only one cluster, a dendrogram can be constructed as shown in Figure 4.20(a). Then, by partitioning this dendrogram at a particular height, one can get the individual clusters. The criteria for evaluating a partition could be similar to what a normalized cut tries to optimize. There are several choices for creating partitions in the dendrogram, and one has to exhaustively compute the objective function value for each partition and choose the optimal one. For example, the detected outlier subsequences are shown in Figure 4.20(b) by selecting a threshold of 5.5 for the height manually. As the figure shows, there are some false alarms and misses in the detected outlier subsequences as the threshold was chosen manually.

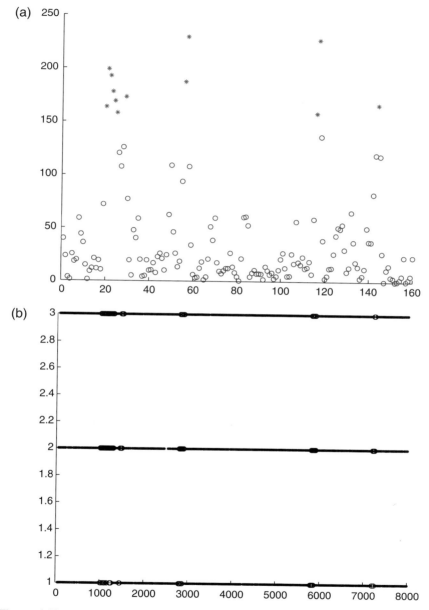

Figure 4.19 Performance of k-means on synthetic time series for case 2. (a) X-axis for time index of context model, Y-axis for cluster indicator value. (b) X-axis for time, Y-axis for symbol.

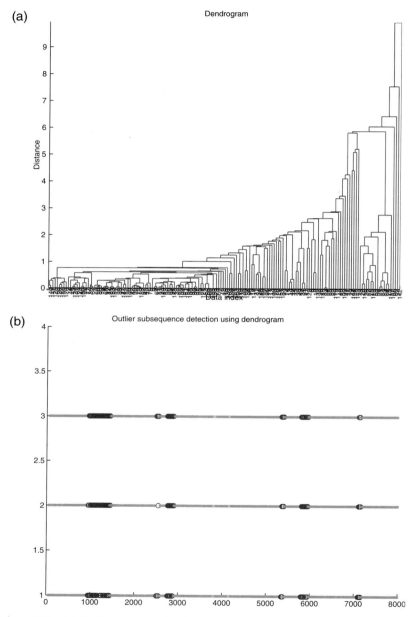

Figure 4.20   (a) Performance of dendrogram cut on synthetic time series for case 2. (b) X-axis for time, Y-axis for symbol.

4.4.8.3   Clustering Based on Factorization of the Affinity Matrix

As mentioned earlier, minimizing N_{cut} minimizes the similarity between groups while maximizing the association within the individual groups. Perona and Freeman [61] modified the objective function of the normalized cut to discover a "salient" foreground object from an unstructured background as shown in Figure 4.21. Since the background is assumed to be unstructured, the objective function of the normalized cut was modified as follows:

$$N_{cut}^*(A, B) = \frac{cut(A, B)}{asso(A, V)} \tag{4.4.9}$$

where cluster A is the foreground and cluster B is the background. Note that the objective function only emphasizes the compactness of the foreground cluster while minimizing the similarity between cluster A and cluster B. Perona and Freeman [61] solve this optimization problem by setting up the problem in the same way as in the case of a normalized cut. The steps of the resulting "foreground cut" algorithm are as follows [61]:

- Calculate the left singular matrix, U, of the affinity matrix, A; $A = USV$.
- Compute the vector $u = SU\mathbf{1}$.
- Determine the index k of the maximum entry of u.
- Define the foreground vector $\mathbf{x}$ as the k_{th} column of U.
- Threshold $\mathbf{x}$, to obtain the foreground and background. $\mathbf{x}$ is similar to the cluster indicator vector in normalized cut.

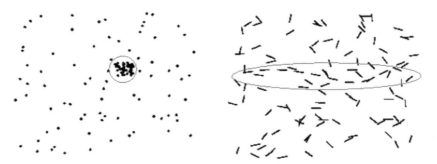

Figure 4.21   Structured "salient" foreground in unstructured background.

The threshold in the last step can be obtained in a similar way as it was obtained for the normalized cut.

For the problem at hand, the situation is reversed (i.e., the background is structured while the foreground can be unstructured). Therefore, the same foreground cut solution should apply as the modified objective function is similar to that of the foreground cut:

$$N_{cut}^{**}(A, B) = \frac{cut(A, B)}{asso(B, V)} \tag{4.4.10}$$

However, a careful examination of the modified objective function would reveal that the term in the denominator $asso(B, V)$ would not be affected drastically by changing the cluster members of A. This is because the background cluster is assumed to be dominant. Hence, minimizing this objective function would be the same as minimizing the value $cut(A, B)$. Minimizing $cut(A, B)$ alone is notorious for producing isolated small clusters. Our experiments with the synthetic time series data also support these observations. Figures 4.22(a) shows the value of objective function $cut(A, B)$ for candidate threshold values in the range of values of the vector $\mathbf{x}$. Figure 4.22(b) shows the value of objective function $N_{cut}^{**}(A, B)$ for the same candidate threshold values. Figures 4.22(c) and (d) show the detected outlier subsequences for the optimal threshold. There are some misses because the modified normalized cut finds isolated small clusters. Note that this procedure could be repeated recursively on the detected background until some stopping criterion is met. For example, the stopping criterion could either be based on percentage of foreground points or on the radius of the background cluster.

As shown in this section, all of the competing clustering approaches need a threshold to be set for detecting outlier subsequences. The alphabet-constrained k-means algorithm needs the knowledge of a number of clusters and a threshold on the difference between the average distance and the semi-Hausdorff distance. The dendrogram-based agglomerative clustering algorithm needs a suitable objective function to evaluate and select the partitions. The foreground cut (modified normalized cut) algorithm finds small, isolated clusters and can be recursively repeated on the background until the radius of the background cluster is smaller than a chosen threshold. Therefore, for the case of a single dominant process with outlier subsequences from a single foreground process, the normalized cut outperforms other clustering approaches.

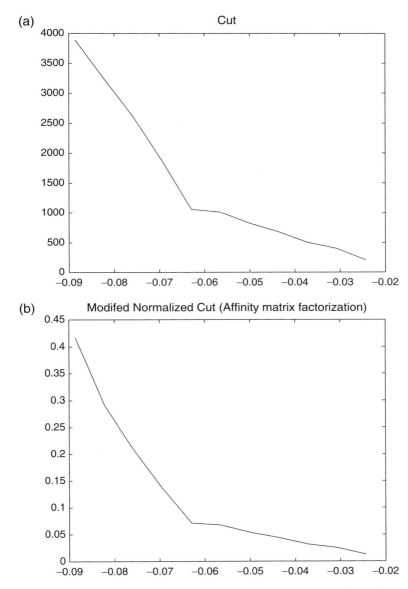

Figure 4.22 Performance of modified normalized cut on synthetic time series for case 2. (a) X-axis for candidate threshold for cut, Y-axis for cut value. (b) X-axis for candidate threshold for modified normalized cut, Y-axis for modified normalized cut value. (c) X-axis for time index of context model, Y-axis for cluster indicator value. (d) X-axis for time, Y-axis for symbol.

(c)

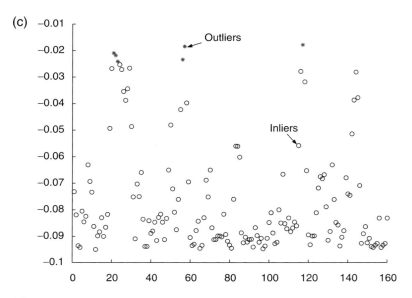

(d) Outlier subsequence detection based on Affinity matrix factorization

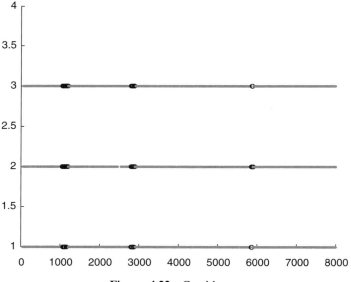

Figure 4.22 Cont'd

In the following section, we consider the next case where there can be multiple foreground processes generating observations against a single dominant background process.

4.4.9 PERFORMANCE OF NORMALIZED CUT FOR CASE 3

The input time series for case 3 is generated using a single dominant background process P_1 and three different foreground processes (P_2, P_3, P_4), and $P(P_1) >> P(P_2) + P(P_3) + P(P_4)$. $P(P_1)$ was set to be 0.8 as in case 2. Figure 4.23(a) shows the input time series. As mentioned earlier, since the normalized cut emphasizes the association between the cluster members for the two clusters resulting from the partition, there are false alarms from the process P_1 in the cluster containing outliers. Figure 4.23(b) shows the normalized cut value for candidate threshold values. There are two minima in the objective function, but the global minimum corresponds to the threshold that results in an outlier cluster with false alarms. Figure 4.23(c) shows the partition corresponding to the global minimum threshold. On the other hand, when the modified normalized cut (foreground cut) is applied to the same input time series, it detects the outliers without any false alarms as shown in Figure 4.23(d) as the objective function does not emphasize the association between the foreground processes.

4.4.9.1 Hierarchical Clustering Using Normalized Cut for Case 4

From the experiments on synthetic time series for case 2 and case 3, we can make the following observations:

- The normalized cut solution is good for detecting distinct time series clusters (backgrounds) as the threshold for partitioning is selected automatically.
- The foreground cut solution is good for detecting outlier subsequences from different foreground processes that occur against a single background.

Both of these observations lead us to a hybrid solution, which uses both the normalized cut and the foreground cut for handling the more general situation in case 4. In case 4, there is no single dominant background process, and the outlier subsequences are from different foreground processes. Figure 4.24(a) shows the input time series for case 4. There are two background processes and three foreground processes.

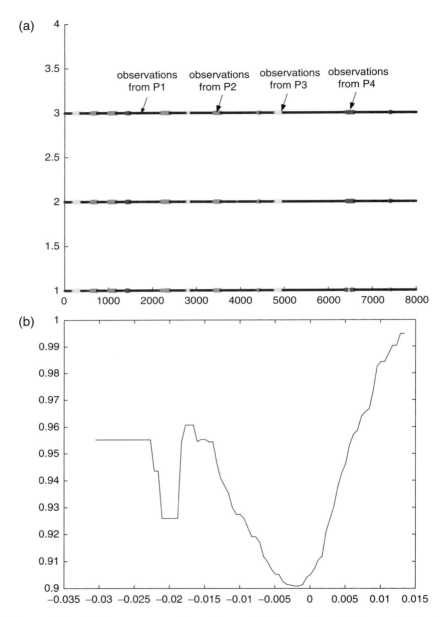

Figure 4.23 Performance comparison of normalized cut and modified normalized cut on synthetic time series for case 3. (a) X-axis for time, Y-axis for symbol. (b) X-axis for candidate threshold for normalized cut, Y-axis for normalized cut value. (c) X-axis for time index of context model, Y-axis for cluster indicator value (normalized cut). (d) X-axis for time index of context model, Y-axis for cluster indicator value (foreground cut).

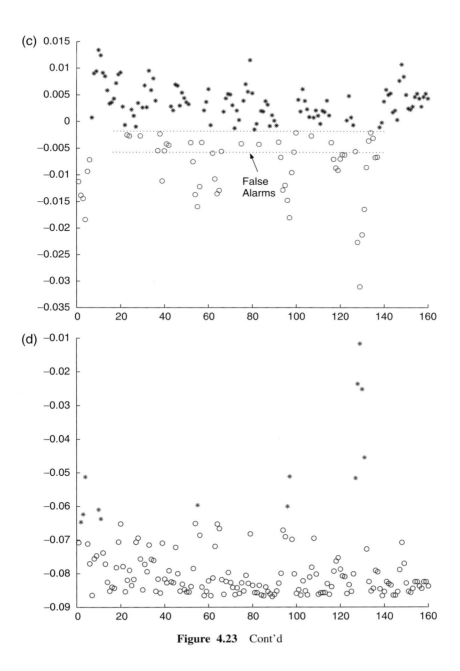

Figure 4.23 Cont'd

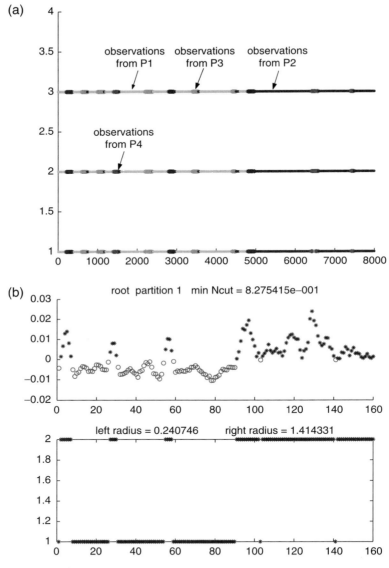

Figure 4.24 Performance of hybrid (normalized cut and foreground cut) approach on synthetic time series for case 4. (a) X-axis for time, Y-axis for symbol. (b) (*top*) X-axis for time index of context model, Y-axis for cluster indicator; (*bottom*) corresponding temporal segmentation. (c) (*top*) X-axis for time index of context model, Y-axis for cluster indicator (second normalized cut); (*bottom*) corresponding temporal segmentation. (d) Final temporal segmentation.

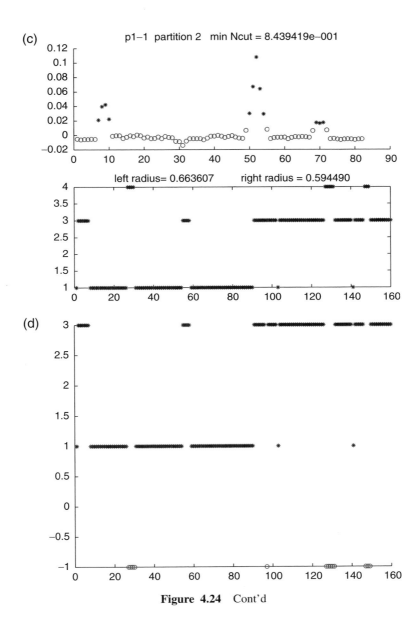

Figure 4.24 Cont'd

Given this input time series and the specifications of a single background in terms of its "compactness" and "size relative to the whole time series" and the maximum percentage of outlier subsequences, we use the following algorithm to detect outlier subsequences:

(1) Use the normalized cut recursively to first identify all individual background processes. The decision of whether or not to split a partition further can be automatically determined by computing the stability of the normalized cut as suggested by Shi and Malik [60] or according to the "compactness" and "size" constraint input to the system.

(2) From the detected distinct backgrounds in Step 1, use the foreground cut recursively to detect the outlier subsequences while making sure that the detected percentage of outliers does not exceed the specified limit.

The "compactness" of a cluster can be specified by computing its radius using the pair-wise affinity matrix as given here:

$$r = \max_{1 \le i \le N} \left(A(i, i) - \left(\frac{2}{N} \sum_{j=1}^{N} A(i, j) \right) + \left(\frac{1}{N^2} \sum_{k=1}^{N} \sum_{j=1}^{N} A(k, j) \right) \right)$$

$$(4.4.11)$$

Here, N represents the number of cluster members. The first term represents the self affinity and is equal to 1. The second term represents the average affinity of ith cluster member with others, and the last term is the average affinity between all the members of the cluster. The computed value of r is guaranteed to be between 0 and 1.

For this input time series, we specified the following parameters: compactness of the background in terms of its radius ≤ 0.5, relative size of background with respect to the size of whole time series ≥ 0.35, and the maximum outlier percentage was set to 20%. Figures 4.24(b) and 4.24(c) show the result of the normalized cut and the corresponding temporal segmentation of the input time series. Figure 4.24(d) shows the final detected outlier subsequences using the foreground cut on individual background clusters.

Now that we have shown the effectiveness of outlier subsequence detection on synthetic time series, we will show its performance on the time series obtained from audio data of sports and surveillance content in the following chapters. In the following section, we analyze how the size of the window used for estimating a context model (W_L) determines the confidence on the detected outlier. The confidence measure is then used to rank the detected outliers.

4.5 Ranking Outliers for Summarization

In this section, first, we show that the confidence on the detected outlier subsequences is dependent on the size of W_L. Second, we use the confidence metric to rank the outlier subsequences.

Recall that in the proposed outlier subsequence detection framework, we sample the input time series on a uniform grid of size W_L and estimate the parameters of the background process from the observations within W_L. Then we measure how different it is from other context models. The difference is caused either by the observations from $\mathbf{P_2}$ within W_L or by the variance of the estimate of the background model. If the observed difference between two context models is "significantly higher than allowed" by the variance of the estimate itself, then we are "somewhat confident" that it was due to the corruption of one of the contexts with observations from $\mathbf{P_2}$.

In the following, before we quantify what is "significantly higher than allowed" and what is "somewhat confident" in terms W_L for the two types of background models (context models for discrete time series and context models for continuous time series) that we will be dealing with, we shall review kernel density estimation.

4.5.1 KERNEL DENSITY ESTIMATION

Given a random sample $x_1, x_2, \ldots, x_n$ of n observations of d-dimensional vectors from some unknown density (f) and a kernel (K), an estimate for the true density can be obtained as follows:

$$\hat{f}(x) = \frac{1}{nh^d} \sum_{i=1}^{n} K\left(\frac{x - x_i}{h}\right) \quad (4.5.1)$$

where h is the bandwidth parameter. If we use the mean squared error (MSE) as a measure of efficiency of the density estimate, the trade-off between bias and variance of the estimate can be seen as shown here:

$$MSE = E[\hat{f}(x) - f(x)]^2 = Var(\hat{f}(x)) + Bias(\hat{f}(x))^2 \quad (4.5.2)$$

It has been shown that the bias is proportional to h^2 and the variance is proportional to $n^{-1}h^{-d}$ [62]. Thus, for a fixed bandwidth estimator, one needs to choose a value of h that achieves the optimal trade-off. We use a data-driven bandwidth selection algorithm proposed in Sheather and Jones [63] for the estimation.

In the following subsection, we present our analysis for binomial and multinomial PDF models for contexts in a discrete time series.

4.5.2 CONFIDENCE MEASURE FOR OUTLIERS WITH BINOMIAL AND MULTINOMIAL PDF MODELS FOR THE CONTEXTS

For the background process to be modeled by a binomial or multinomial PDF, the observations have to be discrete. Without loss of generality, let us represent the set of five discrete labels (the alphabet of $\mathbf{P_1}$ and $\mathbf{P_2}$) by $S = \{A, B, C, D, E\}$. Given a context consisting of W_L observations from S, we can estimate the probability of each of the symbols in S using the relative frequency definition of probability.

Let us represent the unbiased estimator for probability of the symbol A as $\hat{p}_A$. $\hat{p}_A$ is a binomial random variable but can be approximated by a Gaussian random variable with mean as p_A and variance as $\sqrt{\frac{p_A(1-p_A)}{W_L}}$ when $W_L \geq 30$.

As mentioned earlier, in the proposed framework we are interested in knowing the confidence interval of the random variable, d, which measures the difference between two estimates of context models. For mathematical tractability, let us consider the Euclidean distance metric between two PDFs, even though it is only a monotonic approximation to a rigorous measure such as the Kullback-Leibler distance:

$$d = \sum_{i \in S} (\hat{p_{i,1}} - \hat{p_{i,2}})^2 \qquad (4.5.3)$$

Here, $\hat{p}_{i,1}$ and $\hat{p}_{i,2}$ represent the estimates for the probability of ith symbol from two different contexts of size W_L. Since $\hat{p}_{i,2}$ and $\hat{p}_{i,2}$ are both Gaussian random variables, d is an χ^2 random variable with n degrees of the freedom where n is the cardinality of the set S.

Now, we can assert with certain probability,

$$P_c = \int_L^U f_{\chi_n^2}(x)\,dx \qquad (4.5.4)$$

that any estimate of $d(\hat{d})$ lies in the interval $[L, U]$. In other words, we can be P_c confident that the difference between two context model estimates outside this interval was caused by the occurrence of $\mathbf{P_2}$ in one of the contexts. Also, we can rank all the outliers using the probability density function of d.

To verify this analysis, we generated two contexts of size W_L from a known binomial or multinomial PDF (assumed to be the background process, see Figure 4.25 and Figure 4.26). Let us represent the models estimated from these two contexts by M_1 and M_2, respectively. Then, we use bootstrapping and kernel density estimation to verify the analysis on PDF of d as shown:

(1) Generate W_L symbols from M_1 and M_2.
(2) Reestimate the model parameters ($\hat{p}_{i,1}$ and $\hat{p}_{i,2}$) based on the generated data and compute the chosen distance metric (d) for comparing two context models.
(3) Repeat Steps 1 and 2, N times.
(4) Use the kernel density estimation to get the PDF of d, $\hat{p}_{i,1}$, and $\hat{p}_{i,2}$.

Figure 4.27(a) shows the estimated PDFs for binomial model parameters for two contexts of same size (W_L). It can be observed that $\hat{p}_{i,1}$ and $\hat{p}_{i,2}$ are Gaussian random variables in accordance with the Demoivre-Laplace theorem. Figure 4.27(b) estimated PDFs of the defined distance metric for different context sizes. One can make the following two observations:

- The PDF of the distance metric is χ^2 with two degrees of freedom in accordance with our analysis.
- The variance of the distance metric decreases as the number of observations within the context increases from 100 to 600.

Figure 4.27(c) shows the PDF estimates for the case of multinomial PDF as a context model with different context sizes (W_L). Here, the PDF estimate for the distance metric is χ^2 with 4 degrees of freedom which is consistent with the number of symbols in the used multinomial PDF model.

These experiments show the dependence of the PDF estimate of the distance metric on the context size, W_L. Hence for a chosen W_L, one can compute the PDF of the distance metric, and any outlier caused by the occurrence of symbols from another process ($\mathbf{P_2}$) would result in a sample from the tail of this PDF. This would let us quantify the "unusualness" of an outlier in terms of its PDF value.

In the next subsection, we perform a similar analysis for HMMs and GMMs as context models.

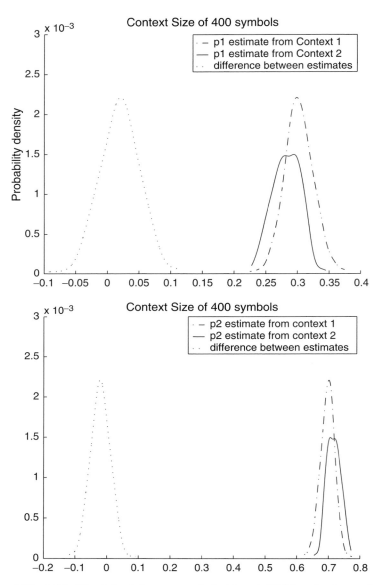

Figure 4.25 PDF of the defined distance metric for binomial PDF as a background model for context sizes of 200 and 400 symbols.

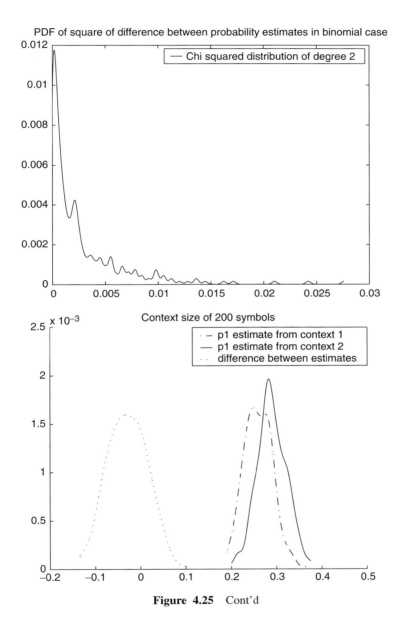

Figure 4.25 Cont'd

(e)

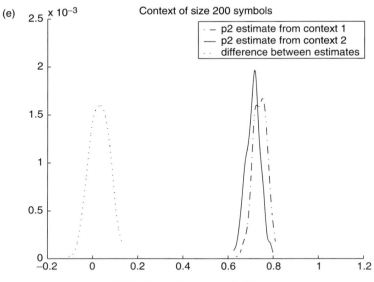

(f)

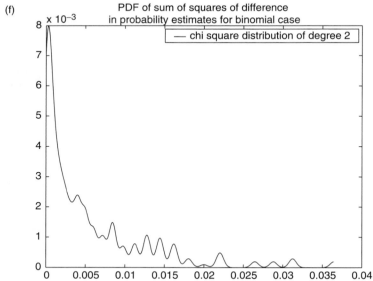

Figure 4.25   Cont'd

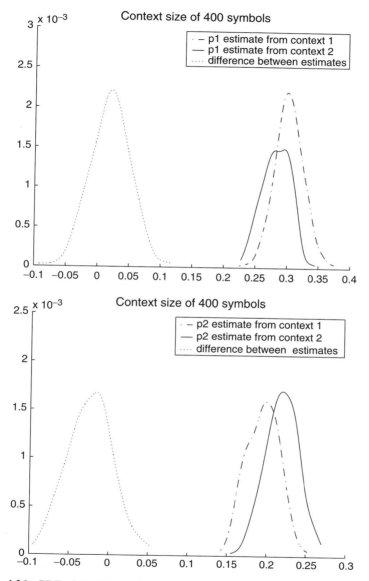

Figure 4.26 PDF of the defined distance metric for multinomial PDF as a background model for a context size of 200.

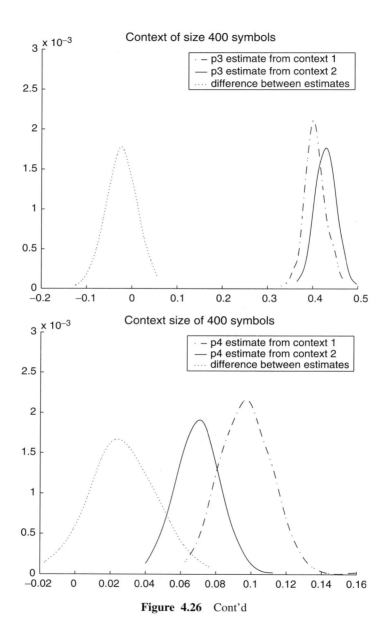

Figure 4.26   Cont'd

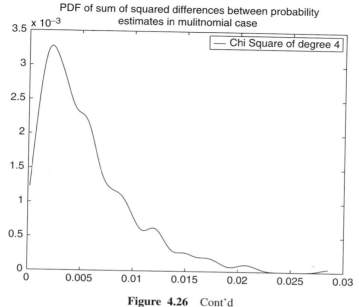

Figure 4.26 Cont'd

4.5.3 CONFIDENCE MEASURE FOR OUTLIERS WITH GMM AND HMM MODELS FOR THE CONTEXTS

When the observations of the memoryless background process are not discrete, one would model its PDF using a Gaussian mixture model (GMM). If the process has first-order memory, one would model its first-order PDF using a hidden Markov model (HMM). Let $\lambda = (A, B, \pi)$ represent the model parameters for both the HMM and GMM where A is the state transition matrix, B is the observation probability distribution, and π is the initial state distribution. For a GMM, A and π are simply equal to 1 and B represents the mixture model for the distribution. For an HMM with continuous observations, B is a mixture model in each of the states. For an HMM with discrete labels as observations, B is a multinomial PDF in each of the states. Two models (HMMs/GMMs) that have different parameters can be statistically equivalent [64] and hence the following distance measure is used to compare two context models (λ_1 and λ_2 with observation sequences O_1 and O_2, respectively).

$$D(\lambda_1, \lambda_2) = \frac{1}{W_L}(\log P(O_1|\lambda_1) + \log P(O_2|\lambda_2)$$

$$- \log P(O_1|\lambda_2) - \log P(O_2|\lambda_1)) \qquad (4.5.5)$$

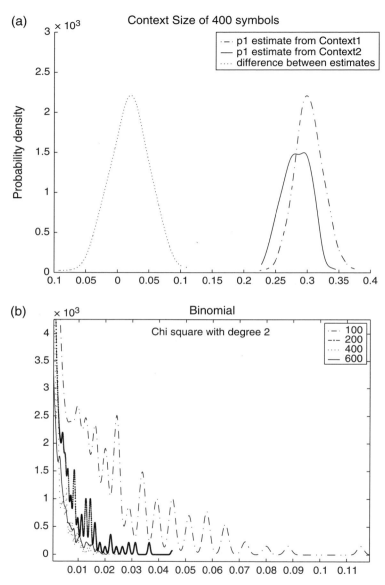

Figure 4.27   PDFs of distance metrics for different background models. (a) PDF of an estimate of a context model parameter. (b) PDF of distances for a binomial context model. (c) PDF of distances for a multinomial context model. (d) PDF of distances for a GMM as a context model. (e) PDF of distances for a HMM as a context model.

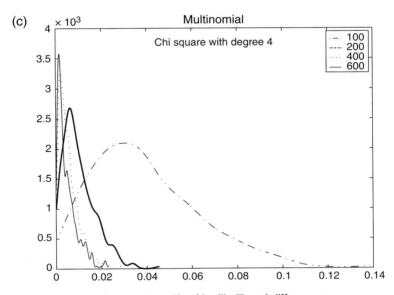

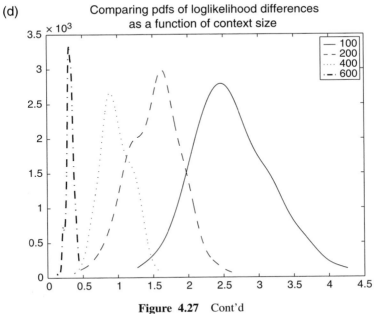

Figure 4.27 Cont'd

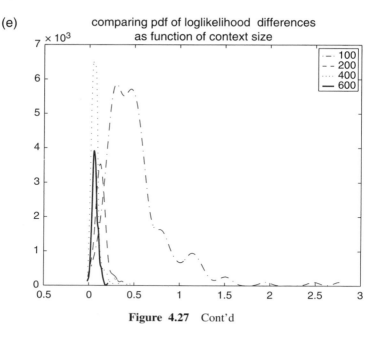

Figure 4.27 Cont'd

The first two terms in the distance metric measure the likelihood of training data given the estimated models. The last two cross terms measure the likelihood of observing O_2 under λ_1 and vice versa. If the two models are different, one would expect the cross terms to be much smaller than the first two terms. Unlike in Section 4.5.2, the PDF of $D(\lambda_1, \lambda_2)$ does not have a convenient parametric form. Therefore, we directly apply bootstrapping to get several observations of the distance metric and use kernel density estimation to get the PDF of the defined distance metric.

Figure 4.27(d) shows the PDF of the log likelihood differences for GMMs for different sizes of context. Note that the PDF gets more skewed or peaky as W_L increases from 100 to 600. The reliability of the two context models for the same background process increases as the amount of training data increases and hence the variance of normalized log likelihood difference decreases. Therefore, again it is possible to quantify the "unusualness" of outliers caused by corruption of observations from another process ($\mathbf{P_2}$). Similar analysis shows the same observations hold for HMMs as context models as well. Figure 4.27(e) shows the PDF of the log likelihood differences for HMMs for different sizes of the context.

4.5.4 *USING CONFIDENCE MEASURES TO RANK OUTLIERS*

In the previous two sections, we looked at the estimation of the PDF of a specific distance metric for comparing context models (memoryless models and HMMs) used in the proposed framework. Then, for a given time series of observations from the two processes ($\mathbf{P_1}$ and $\mathbf{P_2}$), we compute the affinity matrix for a chosen size of W_L for the context model. We use the second generalized eigenvector to detect inliers and outliers. Then the confidence metric for an outlier context M_j is computed as

$$p(M_j \epsilon O) = \frac{1}{\#I} \left(\sum_{i \in I} P_{d,i}(d \le d(M_i, M_j)) \right) \tag{4.5.6}$$

where # is the cordinality operator of a set and $P_{d,i}$ is the density estimate for the distance metric using the observations in the inlier context i. O and I represent the set of outliers and inliers, respectively.

Figure 4.28 shows how two models M_j and M_k estimated from subsequences containing a burst of observations from foreground processes can be ranked based on the PDF of distance between inliers. Since the distance between inlier M_i and M_k ($d(M_i, M_k)$) is more toward the tail of the inlier distance PDF than is the distance between M_i and M_j ($d(M_i, M_j)$), the confidence on M_k being an outlier is higher.

So far in this chapter, we proposed an inlier/outlier–based temporal segmentation of the audio stream of the "unscripted" content. Such a representation is aimed toward postponing content-specific processing to as

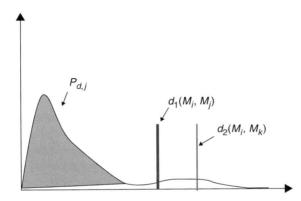

Figure 4.28 Ranking outliers based on PDF of distance between inliers.

late a stage as possible. We described the feature extraction procedures and the audio classification framework that can help represent the input audio either as a continuous time series of low-level audio features or a discrete time series of classification labels.

Then we have developed a time series analysis framework to detect subsequences that are outliers from an input time series. The proposed analysis framework samples subsequences from the input time series and embeds them in a higher dimensional space using a sequence kernel. The sequence kernel is based on statistical models estimated from the subsequences. The kernel matrix obtained after embedding is then used for subsequent clustering and outlier detection using a combination of the normalized cut and the foreground cut. The effectiveness of the proposed framework was demonstrated using synthetic time series. The detected outliers are then ranked based on a confidence measure. The confidence measure is obtained from the PDF of distance between inliers.

In effect, we have addressed the first two questions that we posed in this chapter. In the following section, we apply the time series analysis framework to detect outliers from time series of low-level/midlevel features extracted from sports and situation comedy audio.

4.6   Application to Consumer Video Browsing

In the previous section, we presented the analysis framework to detect outlier subsequences from a time series. Recall that the analysis framework itself was motivated by the observation that "interesting" events in "unscripted" multimedia such as sports and surveillance happen sparsely in the background of usually "uninteresting" events. The outlier detection framework was shown to be effective in detecting statistically unusual subsequences using synthetic time series. In this chapter, we apply the framework for analysis of time series of features/labels extracted from the audio of sports content and situation comedy.

4.6.1   HIGHLIGHTS EXTRACTION FROM SPORTS VIDEO

In the case of sports, our goal is to arrive at an inlier/outlier–based temporal segmentation using the content-adaptive analysis framework. Since the outlier detection is only based on statistical deviation from the usual, not all detected outliers will be semantically "interesting." We need to use domain knowledge at the last stage to prune out outliers that do not correlate with "interesting" highlight moments. Then we systematically collect outliers for the key audio class that correlates with highlight events in sports and train a supervised model. Finally, the acquired domain knowledge in the form of a supervised detector can be used to select outliers that are "interesting." Then, using the ranking one can provide a summary of any desired length to the end user.

In the following sections, we present the results of the proposed framework with sports audio mainly using low-level audio features and semantic audio classification labels at the "8-ms frame-level" and "1-second level." The proposed framework has been tested with a total of 12 hours of soccer, baseball, and golf content from Japanese, American, and Spanish broadcasts.

4.6.1.1 Frame-Level Labels Time Series versus Low-Level Features Time Series

As mentioned earlier, there are three possible choices for time series analysis from which events can be discovered using the proposed outlier subsequence detection framework:

- Low-level MFCC features
- Frame-level audio classification labels
- One-second-level audio classification labels

In this section, we show the pros and cons of using each of these time series for event discovery with some example clips from sports audio. Since the 1-second-level classification label time series is a coarse representation, we can detect commercials as outliers and extract the program segments from the whole video using the proposed framework. For discovering highlight events (for which the time span is only in the order of a few seconds), we use a finer scale time series representation such as the low-level features and frame-level labels.

4.6.1.2 Outlier Subsequence Detection Using One-Second-Level Labels to Extract Program Segments

Based on the observation that commercials are outliers in the background of the whole program at a coarser time scale, we use the 1-second-level audio classification labels as input time series for the proposed framework.

Figure 4.29 shows the affinity matrix for a 3-hour long golf game. We used two-state HMMs as context models with W_L as 120 (W_L) classification labels with a step size of 10 (W_S). The affinity matrix was constructed using the computed pair-wise likelihood distance metric defined earlier. Note that the affinity matrix shows dark regions against a single background. The dark regions, with low affinity values with the rest of the regions (outliers), were verified to be times of occurrences of commercial messages. Since we use the time series of the labels at 1-second resolution, the detected outliers give a coarse segmentation of the whole video into two clusters: the segments that represent the program and the segments that represent the commercials. Also, such a coarse segmentation is possible only because we used a time series of classification labels instead of low-level features. Furthermore, the use of low-level audio features at this stage may bring out some fine scale changes that are not relevant for distinguishing program segments from nonprogram segments. For instance, low-level features may distinguish two different speakers in the content, while a more general speech label would group them as one.

4.6.1.3   Outlier Subsequence Detection from the Extracted Program Segments

Highlight events together with audience reaction in sports video last for only a few seconds. This implies that we cannot look for "interesting" events using the 1-second-level classification labels to extract highlight events. If we use 1-second-level classification labels, the size of W_L has to be small enough to detect events at that resolution. However, our analysis on the confidence measures presented earlier indicates that a small value of W_L would lead to a less reliable context model, thereby producing a lot of false alarms. Therefore, we are left with the following two options:

(1) To detect outlier subsequences from the time series of frame-level classification labels instead of second-level labels
(2) To detect outlier subsequences from the time series of low-level MFCC features

Clearly, using the frame-level classification labels is computationally more efficient. Also, as pointed out earlier, working with labels can suppress irrelevant changes (e.g., speaker changes) in the background process. Figure 4.30(a) shows the cluster indicator vector for a section of golf program segment. The size of W_L used was equal to 8 s of frame level

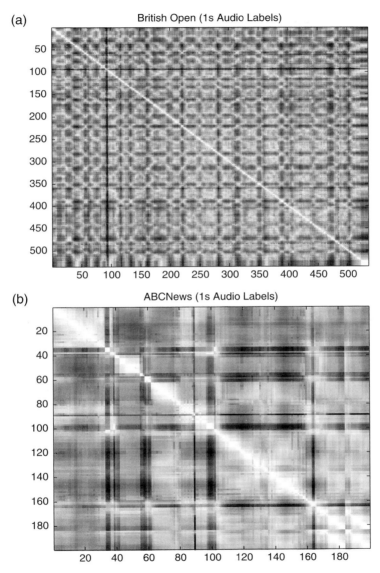

Figure 4.29   (a) Affinity matrix for a 3-hour long British Open golf game and (b) 30-minute news video using one-second-classification labels.

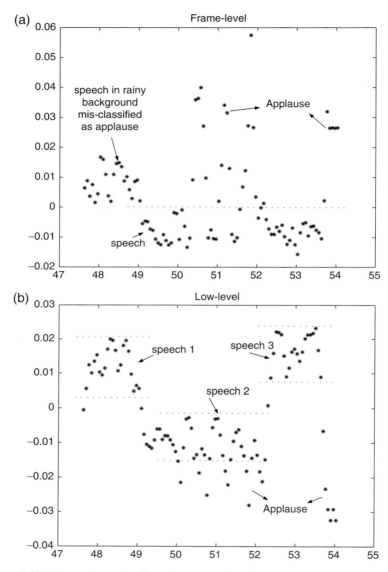

Figure 4.30   Comparison of outlier subsequence detection with low-level audio features and frame-level classification labels for sport and surveillance. (a) Outlier subsequences in frame labels time series for golf. (b) Outlier subsequences in low-level features time series for golf. (c) Outlier subsequences in frame labels time series for soccer. (d) Outlier subsequences in low-level features time series for soccer.

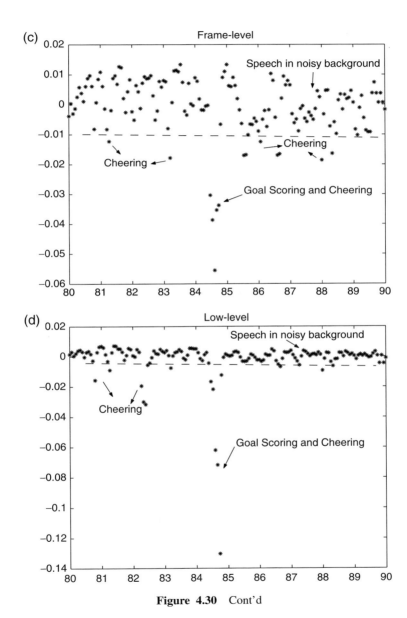

Figure 4.30 Cont'd

classification labels with a step size of 4 s. The context model used for classification labels was a two-state HMM. In the case of low-level features, the size of W_L was equal to 8 s of low-level features with a step size of 4 s (see Figure 4.30(b)). The context model was a two-component GMM. Note that there are outliers at times of occurrences of applause segments in both cases. In the case of outlier detection from low-level features, there were at least two clusters of speech as indicated by the plot of eigenvector and affinity matrix. Speech 3 (marked in the figure) is an interview section where a particular player is being interviewed. Speech 1 is the commentator's speech itself during the game. Since we used low-level features, these time segments appear as different clusters. However, the cluster indicator vector from frame-level labels time series affinity matrix shows a single speech background from the 49th min to the 54th min. However, the outliers from the 47th min to the 49th min in the frame-level time series were caused by misclassification of speech in "windy" background as applause. Note that the low-level feature time series does not have this false alarm. In summary, low-level feature analysis is good only when there is a stationary background process in terms of low-level features. In this example, immobility is lost due to speaker changes. Using a frame-level label time series, on the other hand, is susceptible to noisy classification and can bring out false outliers.

Figures 4.30(c) and 4.30(d) show the outliers in the frame labels time series and the low-level features time series, respectively, for 10 min of a soccer game with the same set of parameters as for the golf game. Note that both of them show the goal scoring moment as an outlier. However, the background model of the low-level features time series has a smaller variance than the background model of the frame labels time series. This is mainly due to the classification errors at the frame levels for soccer audio.

In the next subsection, we present our results on inlier/outlier–based representation for a variety of sports audio content.

4.6.1.4 Inlier/Outlier–Based Representation and Ranking of the Detected Outliers

In this section, we show the results of the outlier detection and ranking of the detected outliers. For all the experiments in this section, we have detected outliers from the low-level features time series to perform an inlier/outlier–based segmentation of every clip. The parameters of the proposed framework were set as follows:

- Context window size $(W_L) = 8$ sec
- Step size $(W_S) = 4$ sec
- *Framerate* at which MFCC features are extracted $= 125$ frames per second
- Maximum percentage of outliers $= 20\%$
- Compactness constraint on the background $= 0.5$
- Relative time span constraint on the background $= 0.35$ and the context model is a two-component GMM

These were not changed for each genre or clip of video. The first three parameters (W_L, W_S, and *Framerate*) pertain to the affinity matrix computation from the time series for a chosen context model. The fourth parameter (maximum percentage of outliers) is an input to the system for the inlier/outlier–based representation. The system then returns a segmentation with at most the specified maximum percentage of outliers. The fifth and sixth parameters (compactness and relative size) help to define what a background is.

First, we show an example inlier/outlier–based segmentation for a 20-min Japanese baseball clip. In this clip, for the first 6 min of the game the audience was relatively noisy compared to the latter part of the game. There is also a 2-min commercial break between the two parts of the game. Figure 4.31 shows the temporal segmentation of this clip during every step of the analysis using the proposed framework. The top part of Figure 4.31(a) shows the result of the first normalized cut on the affinity matrix. The bottom part of Figure 4.31(a) shows the corresponding time segmentation. Since the compactness constraint is not satisfied by these two partitions, the normalized cut is recursively applied. When the normalized cut is applied for the second time, the commercial segment is detected as an outlier, as shown in Figure 4.31(b). Figure 4.31(c) shows the result of the normalized cut on the other partition. The final segmentation is shown in 4.31(d). The outliers were manually verified to be reasonable. As mentioned earlier, outliers are statistically unusual subsequences and not all of them are interesting. Commercial segments and periods of the game during which the commentator is silent but the audience is cheering are some examples that are statistically unusual and not "interesting." Therefore, after this stage one needs to use a supervised detector such as the excited speech detector to pick out only the "interesting" parts for the summary.

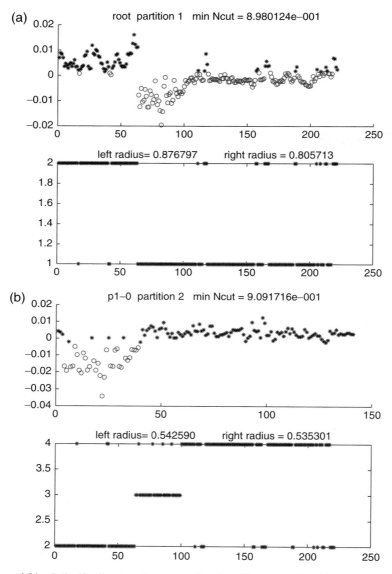

Figure 4.31 Inlier/Outlier–based segmentation for a 20-minute clip of Japanese baseball content. (a) First normalized cut. (b) Second normalized cut. (c) Third normalized cut. (d) Final temporal segmentation after foreground cut from each background.

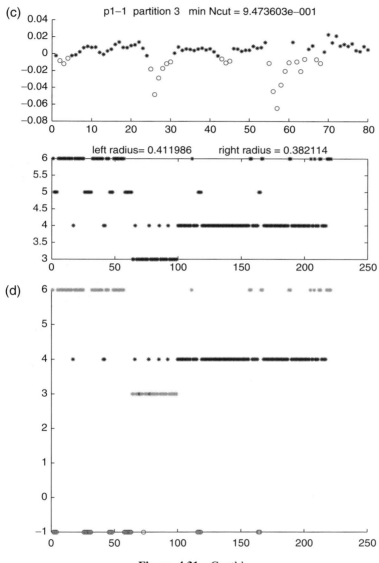

Figure 4.31   Cont'd

4.6.1.5   Results with Baseball Audio

We repeated this kind of inlier/outlier–based segmentation on a total of 4 hours of baseball audio from five different games (two baseball games from Japanese broadcasts and three from American broadcasts). We listened to every outlier clip and classified it by hand as one of the types shown in Table 4.1. Apart from the three types of outliers mentioned earlier, we had outliers when there was an announcement in the stadium and when there was a small percentage of speech in the whole context. In Table 4.1, we also show the average normalized ranking and average normalized distance from the inliers for each type of outlier over all the clips analyzed. It is intuitively satisfying that the speech-with-cheering class is closest to the inliers and has the smallest average rank of all the types. Of all the types, the excited speech-with-cheering and the cheering classes are the most indicative of highlight events.

4.6.1.6   Results with Soccer Audio

With the same setting of parameters, we segmented a total of 6 hours of soccer audio from seven different soccer games (three from Japanese broadcasts, three from American broadcasts, and one from a Spanish broadcast). The types of outliers in the soccer games were similar to those obtained from the baseball games. The results of ranking are also presented for these

Table 4.1   **Outlier ranks in baseball audio; R_1: Average normalized rank using PDF estimate; R_2: Average normalized distance.**

Type of Outlier	R_1	R_2
Speech with cheering	0.3648	0.1113
Cheering	0.7641	0.3852
Excited speech with cheering	0.5190	0.1966
Speech with music	0.6794	0.3562
Whistle, drums with cheering	0.6351	0.2064
Announcement	0.5972	0.3115

types of outliers in Table 4.2. Again, the speech-with-cheering outlier is ranked lowest.

4.6.1.7 Results with Golf Audio

We also segmented 90 minutes of a golf game using the proposed approach. Since the audio characteristics of a golf game are different from those of baseball and soccer, the types of outliers were also different. Applause segments were outliers as expected. The other new types of outliers in golf were when the commentator was silent and when a new speaker was being interviewed by the commentator. The ranks of the detected outlier types are shown in Table 4.3.

4.6.1.8 Results with Public Domain MPEG-7 Soccer Audio Data

In this section, we present the results of inlier/outlier–based segmentation on a public domain data set to allow for comparison with other outlier detection-based highlights extraction systems for further research.

The parameters of the proposed framework were as follows:

- Context window size $(W_L) = 4$ sec
- Step size $(W_S) = 2$ sec

Table 4.2 **Outlier ranks in soccer audio; R_1: Average normalized rank using PDF estimate; R_2: Average normalized distance.**

Type of Outlier	R_1	R_2
Speech with cheering	0.3148	0.1606
Cheering	0.7417	0.4671
Excited speech with cheering	0.4631	0.2712
Speech with music	0.5098	0.2225
Whistle, drums with cheering	0.4105	0.2430
Announcement	0.5518	0.3626

Table 4.3   **Outlier ranks in golf audio; R_1: Average normalized rank using PDF estimate; R_2: Average normalized distance.**

Type of Outlier	R_1	R_2
Silence	0.7573	0.5529
Applause	0.7098	0.4513
Interview	0.1894	0.1183
Speech	0.3379	0.3045

- *Framerate* at which MFCC features are extracted $= 125$ frames per second
- Maximum percentage of outliers $= 20\%$
- Compactness constraint on the background $= 0.5$
- Relative time span constraint on the background $= 0.35$ and the context model is a two-component GMM

The soccer audio for "Foot1002.mpg" was resampled to 16 kHz and inlier/outlier–based segmentation was performed for three clips from 4 min to 19 min, 19 min to 36 min, and 36 min to 52 min. Figure 4.32 shows the affinity matrices and their corresponding temporal segmentation into inliers/outliers for these three clips.

Each of the detected outliers for the Foot1002 clip was listened to, and the segmentation was compared with a ground truth segmentation. The comprehensive ground truth for the whole clip in terms of "remarkable" and "nonremarkable" events is shown in the Table 4.4.

All the "remarkable" events from the ground truth set were part of the detected outliers. There were also other outliers during non-interesting periods of the game, which are statistically unusual but not "interesting." These false alarms occurred in the "non-remarkable" segments of the clip.

4.6.1.9   Browsing System Using the Inlier/Outlier– Based Segmentation

In the previous subsection, we looked at the results of outlier subsequence detection using the low-level features time series for three kinds of sports

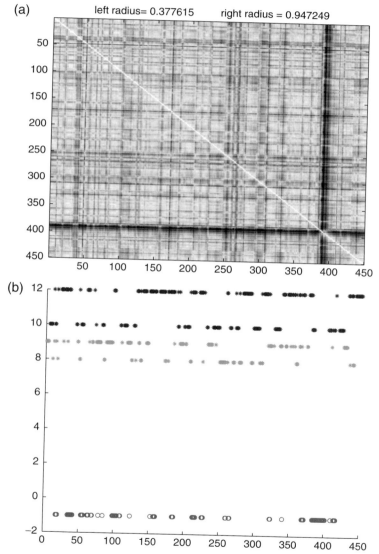

Figure 4.32 Affinity matrices and their corresponding temporal segmentation for the three clips of Foot1002 clip of MPEG-7. (a, b) −4 minutes to 19 minutes. (c, d) −19 minutes to 36 minutes. (e, f) −36 minutes to 52 minutes.

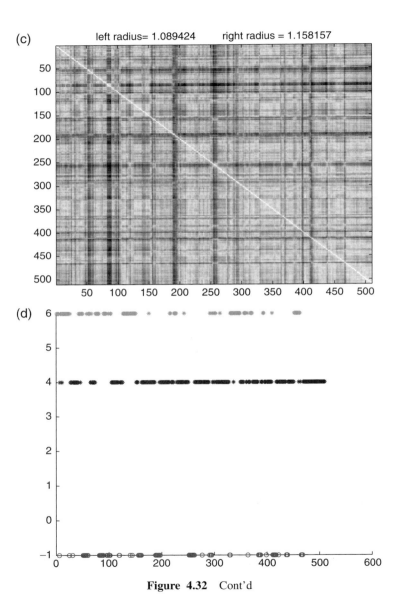

Figure 4.32   Cont'd

(e)

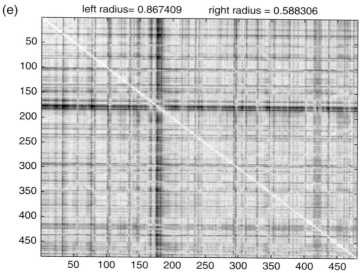

(f)

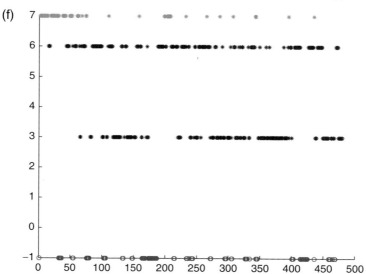

Figure 4.32 Cont'd

Table 4.4   **Ground truth of highlight moments for Foot1002 MPEG-7 clip.**

Start Time	End Time	Event	Start Time	End Time	Event
4:00	4:36	Nothing remarkable	17:18	22:22	Nothing remarkable
4:36	4:40	Attack	22:22	22:34	Free kick
4:40	5:09	Nothing remarkable	22:34	24:20	Nothing remarkable
5:09	5:13	Attack	24:20	24:24	Attack
5:13	5:18	Nothing remarkable	24:24	27:30	Nothing remarkable
5:18	5:22	Attack	27:30	27:40	Attack
5:22	5:48	Nothing remarkable	27:40	31:48	Nothing remarkable
5:48	5:52	Attack	31:48	31:54	Attack
5:52	6:07	Nothing remarkable	31:54	32:40	Nothing remarkable
6:07	6:11	Attack	32:40	32:50	Goal
6:11	7:23	Nothing remarkable	32:50	34:28	Nothing remarkable
7:23	7:28	Attack	34:28	34:34	Attack
7:28	9:15	Nothing remarkable	34:34	41:50	Nothing remarkable
9:15	9:19	Attack	41:50	42:08	Goal
9:19	11:32	Nothing remarkable	42:08	46:56	Nothing remarkable
11:32	11:35	Free kick	46:56	47:01	Free kick
11:35	16:17	Nothing remarkable	47:01	49:24	Nothing remarkable
16:17	16:24	Attack	49:24	49:28	Attack
16:24	16:51	Nothing remarkable	49:24	52:00	Nothing remarkable
16:51	17:18	Goal			

content. Figure 4.33 shows the browsing system based on inlier/outlier–based segmentation for the second half of a soccer game. By using the up and down arrows, the end user can load different temporal segmentations of the content. By using the left and right arrows, the end user can skip to the next detected outlier and start watching the content from there. The system gives the end user feedback as to where he or she is in the content by using an arrow that points to the current media portion in the inlier/outlier segmentation panel.

4.6.2 SCENE SEGMENTATION FOR SITUATION COMEDY VIDEOS

Unlike sports videos, situation comedy videos are carefully produced according to a script. As we saw in the earlier subsections, since sports videos are unscripted they could be summarized by detecting "interesting" events as outliers. Since situation comedy videos are scripted and are composed of a sequence of semantic units (scenes), they cannot be summarized by outlier detection alone. Summarization of situation comedy videos

Figure 4.33   Inlier/Outlier–based segmentation for second half of World Cup final soccer game.

involves detecting the sequence of semantic units (scenes). Then a summary can be provided by allowing a traversal through each of the detected semantic units. Past work on scene segmentation for situation comedy content has relied on a mosaic-based representation of the physical location of a scene [65]. Then an episode is organized in terms of events/scenes happening at a particular physical location. In this section, we apply the same analysis framework that was applied for inlier/outlier–based segmentation of sports content for the task of scene segmentation of situation comedy videos. We are motivated to apply the outlier subsequence detection framework on this content based on the observation that a music clip is played at the end of every scene in the situation comedy content.

We applied the outlier subsequence detection framework to six episodes of situation comedy content with the same setting of parameters that we used for the inlier/outlier segmentation of sports content. The input time series used for segmentation was the low-level features time series. Figure 4.42 (presented later) shows the inlier/outlier–based segmentation of one episode of a situation comedy using the proposed framework. It was observed that all the detected outliers were of the following three types:

- *Type 1*. Burst of laughter from the audience for a comical event
- *Type 2*. Small music clip played during the transition to the next scene
- *Type 3*. Speech of a new character in the episode

Type 1 outlier lets the end user traverse the episode by stepping through comical events marked by a burst of laughter/applause from the audience. Type 2 outlier lets the end user traverse the episode by stepping through the sequence of scenes (semantic units). Type 3 outlier sometimes indicates the theme of the episode and gives the end user an idea of what the current episode is about.

Figure 4.34 shows the detected outliers for six episodes of the situation comedy *Friends* using the foreground cut algorithm on the affinity matrix repeatedly until the outlier budget of 20% has been met.

So far in this section, we applied the proposed content-adaptive analysis framework for inlier/outlier–based segmentation of two types of consumer video: sports content and situation comedy. We showed that the proposed framework is effective in detecting "interesting" moments in baseball, soccer, and golf audio as outliers from the background process. We compared the performance of low-level features time series with the performance of

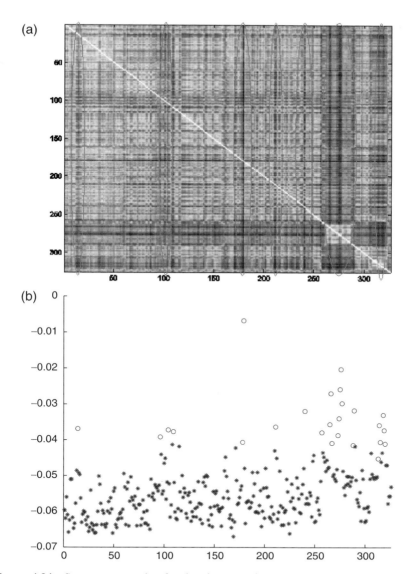

Figure 4.34   Scene segmentation for situation comedy content using the proposed time series outlier subsequence detection framework.

(c)

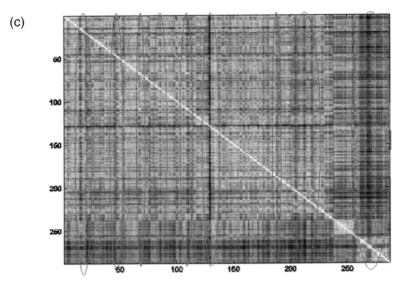

(d)

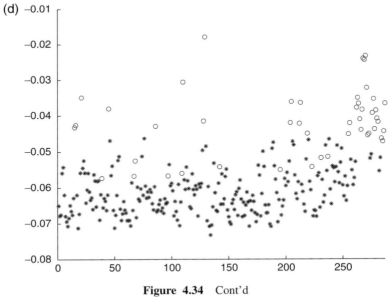

Figure 4.34   Cont'd

(e)

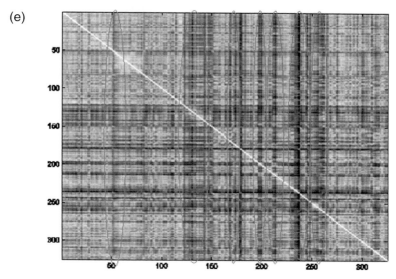

(f)

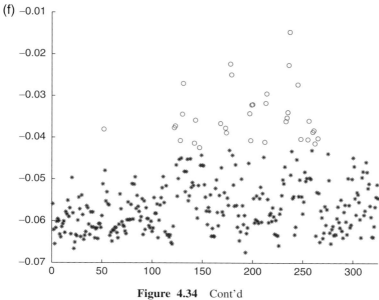

Figure 4.34 Cont'd

(g)

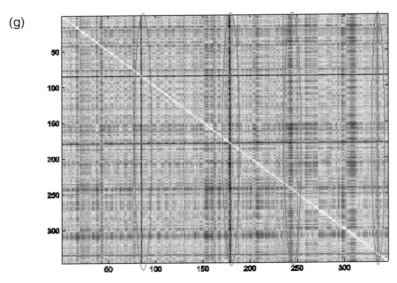

(h)

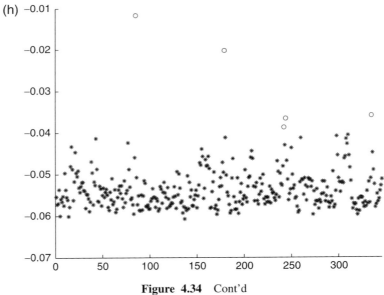

Figure 4.34 Cont'd

(i)

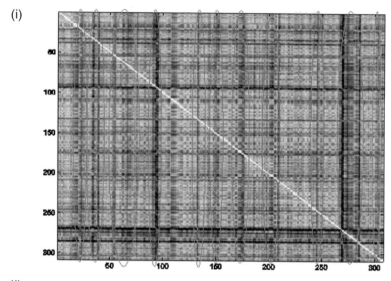

(j)

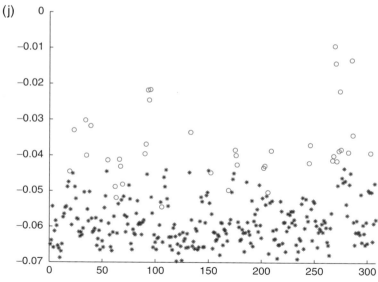

Figure 4.34 Cont'd

(k)

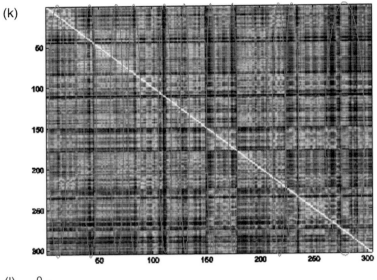

(l)

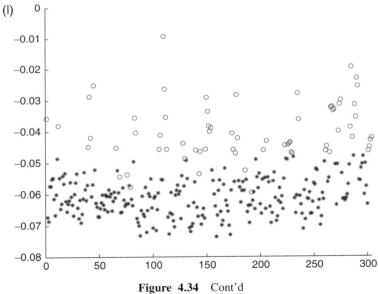

Figure 4.34   Cont'd

discrete classification labels time series for the task of inlier/outlier segmentation. Finally, we showed the effectiveness of the proposed framework for scene segmentation of situation comedy content.

4.7 Systematic Acquisition of Key Audio Classes

We have so far applied the proposed analysis framework to sports audio to perform inlier/outlier–based segmentation of the content and also verified that "interesting" moments in sports audio are a subset of all the detected outliers. Since the proposed framework detects outliers based on statistical deviation from the usual, not all detected outliers will correspond to events of interest. Therefore, to provide an "interesting" summary to the end user, recall that the proposed framework brings in domain knowledge at the last stage to prune out outliers that are not "interesting." In the context of audio analysis, bringing in domain knowledge simply means having a detector for a particular class of audio outlier that indicates the event of interest.

In this section, we show how we can acquire such key audio classes in a systematic way for sports and surveillance content. In domains such as sports, one can also choose the key audio class based on intuition or some partial domain knowledge. We show that the precision-recall performance of the key audio class that is chosen systematically using the proposed framework is superior to the performance of the key audio class chosen based on intuition. In other domains (such as surveillance, which has a large volume of data and no domain knowledge), acquiring domain knowledge systematically becomes even more important. We apply the time series analysis framework to elevator surveillance audio data to discover key audio classes systematically.

4.7.1 APPLICATION TO SPORTS HIGHLIGHTS EXTRACTION

From all the experiments with sports audio content presented in the previous chapter, one can infer that the proposed framework not only gave an inlier/outlier–based temporal segmentation of the content but also distinguishable sound classes for the chosen low-level features in terms of distinct backgrounds and outlier, sound classes. Then, by examining individual clusters and outliers, one can identify consistent patterns in the data that correspond to the events of interest and build supervised statistical learning models. Thus, the proposed analysis and representation framework can be used for the systematic choice of key audio classes as shown in Figure 4.35.

We cite an example in which this framework was useful for acquiring domain knowledge. In the previous chapter, we showed that one can also use audio classification labels as a time series and discover events. However, the choice of audio classes to be trained for the audio classification framework involves knowledge of the domain in terms of coming up with representative sound classes that cover most of the sounds in the domain. For example, we chose the following five classes for the audio classification framework in the sports domain: applause, cheering, music, speech, and speech with music. The intuition was that the first two classes capture the audience reaction sounds and the rest of the classes represent the bulk of sounds in the "uninteresting" parts of the sports content. Such a

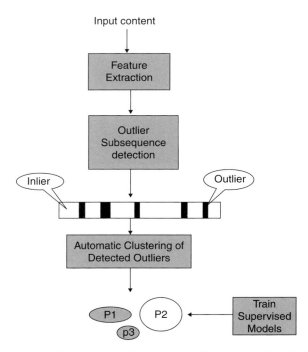

Figure 4.35   Systematic acquisition of domain knowledge using the inlier/outlier–based representation framework.

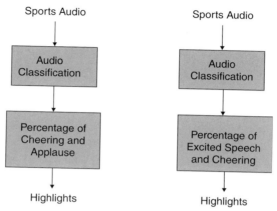

Figure 4.36 Comparison of highlights extraction using audio classification: Applause and cheering versus excited speech with cheering.

classification framework can be used to gauge audience reaction by computing the percentage of audience reaction classification labels (applause and cheering) in a given time segment as shown in Figure 4.36.

Now, let us compare the performance of these two classes (applause and cheering) with the performance of a "highlight" class discovered systematically from the sports audio data. The "highlight" class was discovered by using the proposed outlier subsequence detection framework on the low-level features time series. After clustering the detected outliers and listening to representative members from each outlier cluster, we discovered that the "key" highlight class is a mixture of audience cheering and commentator's excited speech and not the cheering of the audience alone. We collected several training examples from the detected outliers for the key highlight audio class and trained a minimum description length GMM as described in Chapter 3. We use the percentage of this highlight audio class label to rank the time segments in the input sports video as shown in Figure 4.37. In the same figure, we have also shown the ranking for the time segments using percentage of cheering and applause labels. As indicated in Figure 4.37, the highlight audio class reduces false alarms by eliminating segments of the game where only the audience cheers. By choosing the same threshold on these two ranked streams, we select the time segments that have a rank greater than the threshold and verify manually if they were really highlight moments. In terms of precision, the highlights extraction system based on this discovered highlight class outperforms the state-of-the-art

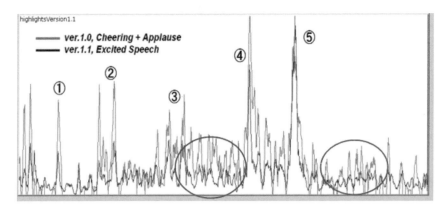

Figure 4.37   Comparison of highlights ranking using cheering and applause against using the "highlight" audio class.

highlights extraction system that uses the percentage of cheering audio class as a measure of interestingness, as shown later in Figure 4.38 [66]. The learned model has been tested for highlights extraction from 27 sports videos including soccer, baseball, sumo wrestling, and horse racing.

Now that we have a highlight class that gives us a better precision-recall performance, we proceed to interpret the meaning of the minimum description length-Gaussian mixture model (MDL-GMM) of this class by inferring what each component is modeling for the given training data set. The MDL solution for the number of components in the GMM for the highlight audio class data set was 4. Given an input feature vector, y_n and a K component GMM with θ as learned parameters, we can calculate the probability that a mixture component, k, generated y_n by using Bayes's rule as given here:

$$p(k/y_n, \theta) = \frac{p(y_n/k, \theta)\pi_k}{\sum_{k=1}^{K} p_{y_n}(y_n \mid k, \theta)\pi_k}$$

Then we can assign y_n to the mixture component for which the posterior probability $(p(k/y_n, \theta))$ is maximum. If we append all the audio frames corresponding to each of the mixture components, we can interpret the semantic meaning of every component [67]. We performed mixture component inferencing for the highlight audio class using the MDL-GMM.

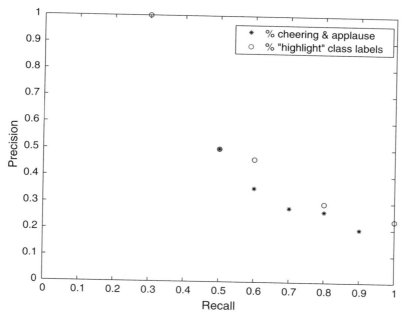

Figure 4.38 Comparison of precision-recall performance using cheering and applause against using the discovered "highlight" audio class.

We find that one of the components predominantly models the excited speech of the commentator, and another component models the cheering of the audience.

In the preceding experiment, we only replaced the key audio class in the audio classification framework from applause, cheering, music, speech, speech with music to applause, cheering, excited speech with cheering, music, speech, and improved the precision-recall performance for sports highlights extraction. The key audio class "excited speech with cheering" was obtained from the data by clustering all the detected outliers from a game. Now we can extend this approach to learning the background sounds for a game from the data and use a simple two-class classification framework with the background sound class, excited speech with cheering, instead of five sound classes to perform highlights extraction. We collect all inliers from the second half of a World Cup soccer game and train an MDL-GMM. This MDL-GMM becomes the background class for this game. Then we cluster the detected outliers from the same clip to collect training data for the foreground class for this game. Then, for a new test clip of this game, we can compute the likelihood of that clip under the trained background model and trained foreground model. Figure 4.39 shows the distribution of 4-sec segments from a 20-min soccer test clip from the same game in the likelihood space of these two classes.

The points marked as "◯" correspond to points for which the likelihood of a foreground model is greater than the likelihood of a background model and were verified to be highlight moments. We had only missed two more "interesting" moments closer to the boundary of background points. Again, the user can be allowed to select the boundary for foreground and background in the likelihood space. Note that the other type of outliers (statistically unusual but semantically "interesting") will have low likelihood in both the models, as shown in Figure 4.39. This experiment shows that with systematic choice of the background class and the foreground class (key audio class), we can alleviate the need for selecting other classes also by intuition.

In other words, the systematic choice of audio classes led to a distinct improvement in the highlights extraction, even though sports is a very familiar or well-known content genre. Note that with less understood

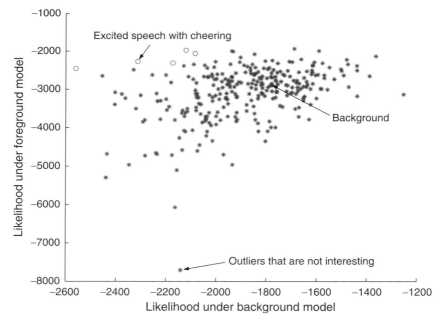

Figure 4.39 Four-second segments of a 20-minute soccer test clip in (foreground-background) likelihood space.

domains such as surveillance, the choice of audio classes based on pure intuition could lead to even worse accuracy of event detection. Furthermore, for surveillance domains especially, the audio classes cannot all be anticipated since there is no restriction on the kinds of sounds.

In the following section, we use this framework for selecting the sound classes to characterize the elevator surveillance audio data and achieve accurate detection of notable events. In this case, the isolation of the elevator car results in a relatively noise-free environment, which makes the data set much more amenable to analysis than is broadcast sports content.

4.7.2 EVENT DETECTION IN ELEVATOR SURVEILLANCE AUDIO

In this section, we apply the proposed analysis to a collection of elevator surveillance audio data for systematic acquisition of key audio classes for this domain. The data set contains recordings of suspicious activities in elevators as well as some event-free clips. Since most of the suspicious events are outliers in a background of usual events, we are motivated to apply the proposed outlier subsequence detection framework for the task of inlier/outlier–based segmentation of the surveillance content. Then, by examining the detected outliers from the suspicious clips, we can systematically select key audio classes that indicate suspicious events. A list of key audio classes is shown in Table 4.5. By examining inliers and outliers from the segmentation of event free clips, we can identify sound classes that characterize usual events in the elevator.

The elevator surveillance audio data consists of 126 clips (2 hours of content) with suspicious events and 4 clips (40 minutes of content) that are without events. We extract low-level features from 61 clips (1 hour of content) of all the suspicious event clips and 4 clips of normal activity in the elevators (40 minutes of content). Then for each of the clips we perform inlier/outlier–based segmentation with the proposed framework to detect outlier subsequences. A subsequent clustering of the detected outliers will help us discover key audio classes that correlate with the event of interest.

The parameters of the proposed framework were set as follows:

- Context window size (W_L) = 4 sec
- Step size (W_S) = 2 sec
- *Framerate* at which MFCC features are extracted = 125 frames per second
- Maximum percentage of outliers = 30%

Table 4.5   **Systematically acquired sound classes from detected outliers and inliers after analysis of surveillance content from elevators.**

Label	Sound Class
1	Alarm
2	Banging
3	Elevator background
4	Door opening and closing
5	Elevator bell
6	Footsteps
7	Nonneutral speech
8	Normal speech

- Compactness constraint on the background $= 0.5$
- Relative time span constraint on the background $= 0.35$ and the context model is a two-component GMM

A two-component GMM was used to model the PDF of the low-level audio features in the 4 s context window. Figures 4.40(a) to (g) show a hierarchical segmentation result of an event-free surveillance audio clip using the proposed time series analysis framework. The first normalized cut solution of the affinity matrix in Figure 4.40(b) is shown in Figure 4.40(a). It was verified that one of the clusters corresponds to time segments during which the elevator was not used. The other cluster corresponds to segments during which the elevator was in use. Further splitting this cluster, as shown in Figures 4.40(c) through 4.40(e), finds the following three types of outliers:

- When there are sounds of footsteps
- When there are screeching sounds due to the elevator door opening or closing
- When there are conversations within the elevator

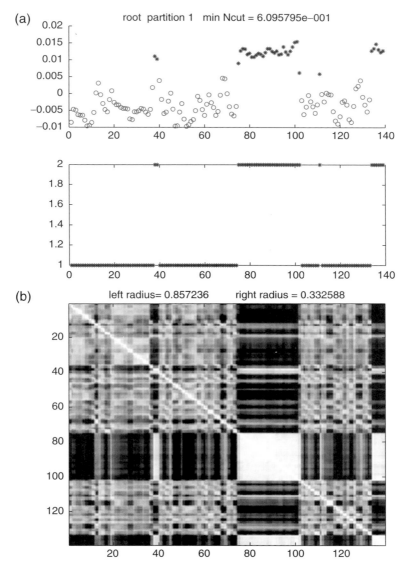

Figure 4.40 Example hierarchical segmentation for an event-free clip in elevator surveillance content.

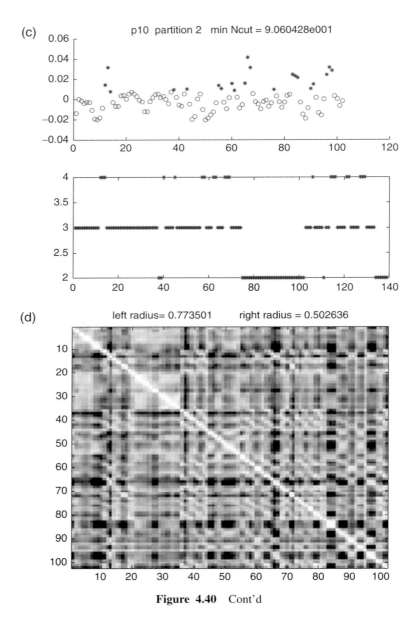

Figure 4.40   Cont'd

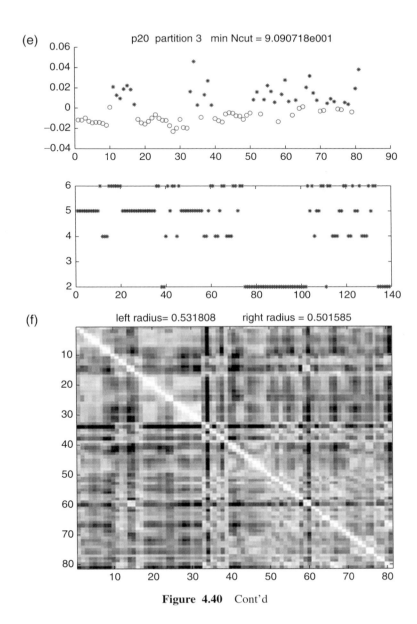

Figure 4.40   Cont'd

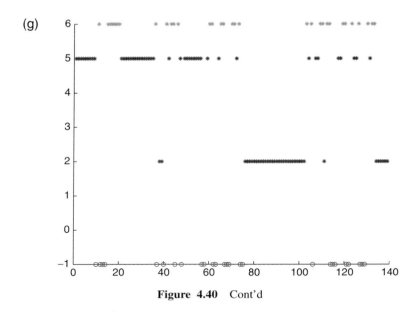

Figure 4.40   Cont'd

These outliers are consistent patterns detected by applying the analysis framework to event-free clips. This implies that we can characterize the usual elevator activity by training a supervised detector for the following four sound classes: elevator background (i.e., inliers from event-free clips), footsteps, normal speech within the elevator, and elevator door opening and closing sounds.

To characterize the sounds during suspicious events in elevators, we carry out a similar temporal segmentation for each of the 61 clips. Figures 4.41(a) to (c) show the segmentation result for a clip with suspicious activity. The outliers in this case turned out to be from the following two categories: banging sounds against elevator walls and nonneutral (excited) speech. In all the clips with suspicious activity, the outliers consistently turned out to be clips of banging sounds against elevator walls and excited speech.

Finally, by performing a simple clustering operation on the detected outliers from the analysis of event-free clips and suspicious clips, we collected training data for the following sound classes to characterize the complete surveillance audio data.

Since these sound classes were obtained as distinguishable sounds from the data, we already know what features and supervised learning method

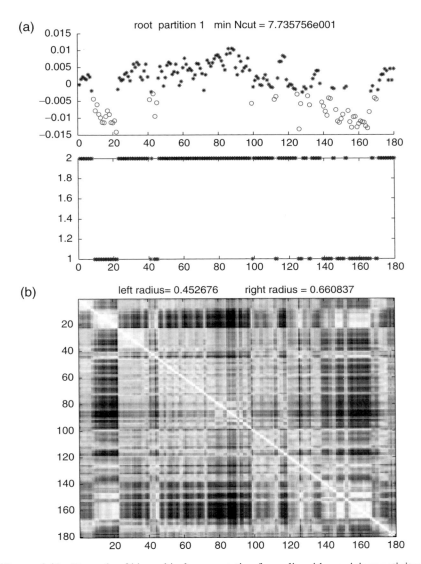

Figure 4.41 Example of hierarchical segmentation for a clip with suspicious activity in elevator surveillance content.

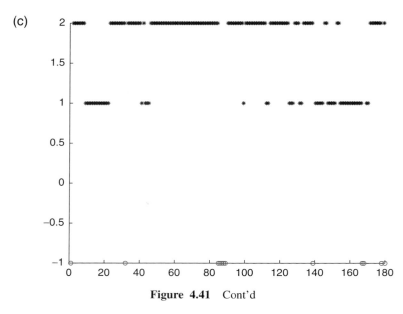

Figure 4.41   Cont'd

are to be used for modeling them. Hence, we use a minimum description length GMM to model the distribution of low-level features. The learned models were used to classify every second of audio from the remaining clips with suspicious activity in elevators. The classification system correctly classified banging sounds and nonneutral speech in all of the suspicious clips thereby enabling detection of these events. Figure 4.42 shows classification results for every second of audio in each of the four types of suspicious activity in elevators. Class labels that indicate suspicious activity (banging sounds and nonneutral speech) are 2 and 8 while other labels characterize usual activities.

In this chapter, we showed two applications of the systematic acquisition of key audio classes. In the case of sports audio, we showed that the key audio class systematically acquired outperforms the audio class chosen based on intuition. In the case of elevator surveillance audio data, we successfully characterized the domain by finding key audio classes without any a priori knowledge to detect suspicious activity.

4.8   Possibilities for Future Research

We proposed a content-adaptive analysis and representation framework for audio event discovery from unscripted multimedia. The proposed

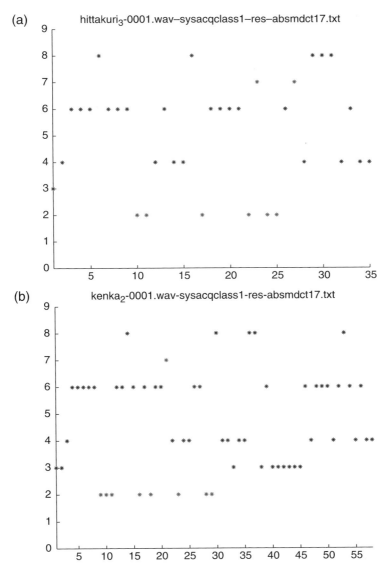

Figure 4.42   Classification results on test clips with suspicious activity.

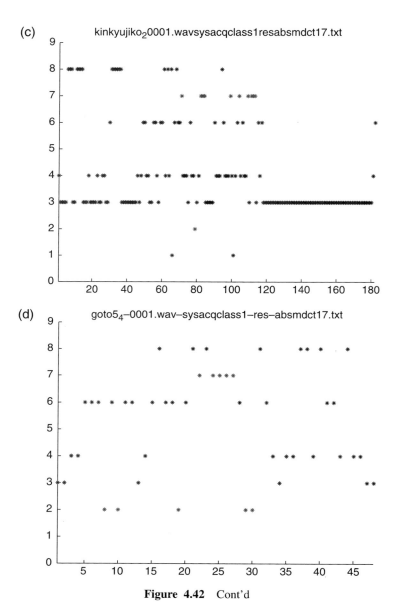

Figure 4.42 Cont'd

framework is based on the observation that "interesting" events happen sparsely in a background of usual events. We used three time series for audio event discovery: low-level audio features, frame-level audio classification labels, and 1-second-level audio classification. We performed an inlier/outlier–based temporal segmentation of these three time series. The segmentation was based on eigenvector analysis of the affinity matrix obtained from statistical models of the subsequences of the input time series. The detected outliers were also ranked based on deviation from the background process. Experimental results on a total of 12 hours of sports audio from three different genres (soccer, baseball, and golf) from Japanese, American, and Spanish broadcasts show that unusual events can be effectively extracted from such an inlier/outlier–based segmentation resulting from the proposed framework. It was also observed that not all outliers correspond to "highlight" events, and one needs to incorporate domain knowledge in the form of supervised detectors at the last stage to extract highlights. Then, using the ranking of the outliers, a summary of desired length can be generated. We also discussed the pros and cons of using the aforementioned three kinds of time series for audio event discovery. We also showed that unusual events can be detected from surveillance audio without any a priori knowledge using this framework. Finally, we have shown that such an analysis framework resulting in an inlier/outlier–based temporal segmentation of the content postpones the use of content-specific processing to as late a stage as possible and can be used to systematically select the key audio classes that indicate events of interest.

(1) In this chapter, the proposed framework has been applied mainly for summarization by analyzing audio features of the recorded content offline. It can be easily adapted for online event detection in surveillance as well. The background model can be updated constantly and can be used to flag deviations from the usual. To reduce the number of false alarms to a manageable level, one can use the systematically acquired audio classes obtained from offline analysis to prune out outliers that are not "interesting." Figure 4.43 shows an example adaptation for the elevator surveillance application.

The adaptive background estimation is similar to well-known background estimation algorithms used for foreground object segmentation in video. In the case of audio analysis, foreground audio objects are more interesting and closer to semantics.

(2) The proposed content-adaptive time series analysis framework brings out outlier patterns in a time series without any a priori

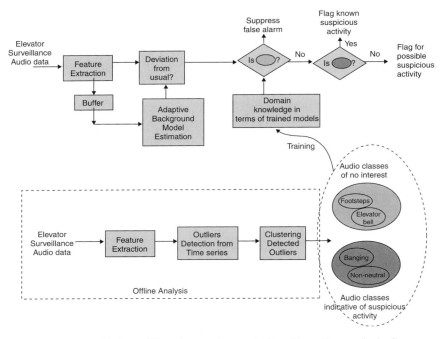

Figure 4.43 Combining offline time series analysis with online analysis for event detection in surveillance.

knowledge. Therefore, it is well suited for domains with a large volume of data that cannot be easily visualized by humans. In such domains, selecting patterns of interest by intuition is not an option. However, by using this framework, one can systematically learn about the domain in terms of what patterns are correlated with the events of interest in the domain. One example domain with such characteristics is the audio data from machines used for detecting faulty machines. Obviously, it is difficult to guess the characteristic sound of a faulty machine beforehand.

(3) The proposed time series can also be carried out using video features extracted from the content. If we use low-level features like a color histogram, the outlier detection will only detect shot changes. However, working with a time series of midlevel semantic labels (e.g., motion activity, face detection result on a frame) may bring out interesting patterns in the content. For instance, a burst of

high motion activity may indicate some event in surveillance content. One can also work with object tracking results from frame to frame to detect anomalous motion patterns [68]. Once we have a meaningful segmentation using video features, a natural question to ask is how can we fuse another segmentation of the same content using audio features? Can we discover cause-and-effect relationships between audio and video outliers?

(4) The proposed time series analysis framework is quite general and is not limited to analyzing time series of audio features. Therefore, the framework is applicable to other domains where "interesting" events can be modeled as outliers from the usual. Computer intrusion detection systems are examples of systems that try to flag unusual activity based on an adaptive model of usual activity. The input time series in such systems is usually a sequence of system calls that represents a user's behavior in that system. By modeling a particular user's normal sequence of system calls, one can detect anomalies triggered by an attack on the computer.

Chapter 5 | Video Indexing

5.1 Introduction

5.1.1 MOTIVATION

As mentioned in Chapter 1, video browsing consists of top-down traversal—that is, summarization—and bottom-up traversal, namely indexing. This chapter focuses on video indexing. The basic video indexing task is to find objects in the search space that match the query object. Therefore the indexing task requires each object to have a consistent description and a similarity metric that helps establish how closely any two descriptions match. In this chapter we focus on descriptions based on low-level video and audio features. Such indexing departs from key-word or text-based indexing in two important ways. First, the matching with low-level features is not necessarily exact. Second, low-level, feature-based descriptions can be automatically generated. Since video indexing is a well-studied topic, we present only an overview here.

5.1.2 OVERVIEW OF MPEG-7

More and more audio-visual information is becoming available in digital form on the World Wide Web as well as other sources. The desire for efficient management of such widespread content has motivated the new MPEG-7 (see Manjunath's MPEG-7 book [69] and Motion Picture Experts Group (MPEG) [88]) standard. It is formally called "Multimedia Content Description Interface" and provides a rich set of standardized tools to describe multimedia content. The standard specifies a standard set of descriptors and description schemes. An MPEG-7 description of content then consists of instantiated description schemes. These descriptions make the content easy to find, retrieve, access, filter, and manage. Consequently, these descriptions enable content-based access such as retrieval from multimedia databases, video browsing, and summarization. Our subsequent section on indexing with low-level features will use MPEG-7 descriptors as representative examples. Since our emphasis has been on motion descriptors, those are presented in greater detail.

The MPEG-7 descriptors have been developed with several criteria in mind. Salient criteria include matching performance, low computational complexity of matching and extraction techniques, compactness, scalability, and invariance to operations such as rescaling, shifting, and rotation.

5.2   Indexing with Low-Level Features: Motion

5.2.1   INTRODUCTION

The motion features of a video sequence constitute an integral part of its spatiotemporal characteristics and are thus indispensable for video indexing. Past work on video indexing using motion characteristics has largely relied on camera motion estimation techniques, trajectory-matching techniques, and aggregated motion vector histogram–based techniques [70–83]. Expressing the motion field, coarse or fine, of a typical video sequence requires a huge volume of information. Hence, the principal objective of motion-based indexing is concisely and effectively encapsulating essential characteristics of the motion field. Such indexing techniques span a wide range of computational complexity, since the motion fields can be sparse or dense, and the processing of the motion fields ranges from simple to complex. Motion feature extraction in the MPEG-1/2 compressed domain has been popular because of the ease of extraction of the motion vectors from the compressed bit stream. However, since compressed domain motion vectors constitute a sparse motion field, they cannot be used for computing descriptors that require dense motion fields. Motion-based indexing is useful by itself since it enables motion-based queries, which is useful in domains such as sports and surveillance in which motion is the dominant feature. For example, the motion activity descriptor can help detect exciting moments from a soccer game, or the motion trajectory descriptor can help distinguish a lob from a volley in a tennis game. Motion-based indexing has also been shown to significantly improve the performance of similarity-based video retrieval systems [79] when combined with other fundamental features such as color, texture, and shape. Motion descriptions can also be the basis for complementary functionalities, such as video hyperlinking based on trajectories [84, 85], refined browsing based on motion [86], or refinement of a table of contents [85, 87].

5.2.2   OVERVIEW OF MPEG-7 MOTION DESCRIPTORS

The MPEG-7 motion descriptors cover the range of complexity and functionality mentioned in the introduction, enabling MPEG-7 to support a broad range of applications. The common guiding principle is to maintain simple extraction, simple matching, concise expression, and effective characterization of motion characteristics. They are organized as illustrated in Figure 5.1.

5.2.3   CAMERA MOTION DESCRIPTOR

5.2.3.1   Salient Syntax and Semantics

This descriptor characterizes three-dimensional (3D) camera motion parameters. It is based on 3D camera motion parameter information, which can be automatically extracted or generated by capture devices. An illustration of the 3D camera motion information is shown in Figure 5.2. The camera motion descriptor supports the following well-known basic camera operations: fixed, panning (horizontal rotation), tracking (horizontal transverse movement, also called traveling in the film industry), tilting (vertical rotation), booming (vertical transverse movement), zooming (change of the focal length), dollying (translation along the optical axis), and rolling (rotation around the optical axis).

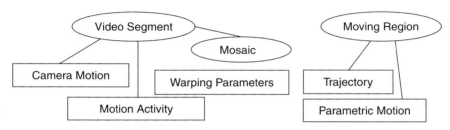

Figure 5.1   MPEG-7 motion descriptions.

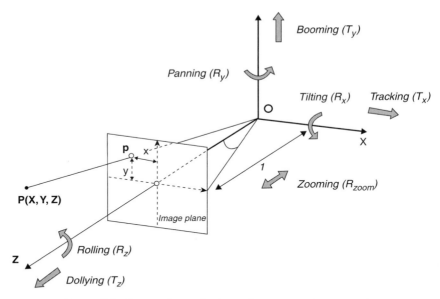

Figure 5.2   Perspective projection and camera motion parameters.

5.2.3.2   Highlights of Extraction

Note that this descriptor can be instantiated by the capture device during capture. If that is not possible, the extraction relies on getting the camera motion parameters from the motion field. The camera motion parameters consist of three camera translation parameters (T_x, T_y, and T_z), three camera rotation parameters (R_x, R_y, and R_z), and one camera zoom parameter (R_{zoom}). The camera motion according to these seven parameters induces an image motion described as follows:

$$u_x = -\frac{f}{Z}(T_x - xT_z) + \frac{xy}{f}R_x - f\left(1 + \frac{x^2}{f^2}\right)R_y$$

$$+ yR_z + ftan^{-1}\left(\frac{x}{f}\right)\left(1 + \frac{x^2}{f^2}\right)R_{zoom} \qquad (5.2.1)$$

$$u_y = -\frac{f}{Z}(T_y - yT_z) - \frac{xy}{f}R_y + f\left(1 + \frac{y^2}{f^2}\right)R_x$$

$$- xR_z + ftan^{-1}\left(\frac{y}{f}\right)\left(1 + \frac{y^2}{f^2}\right)R_{zoom} \qquad (5.2.2)$$

These equations are obtained by using perspective transformation of a physical point $P(X, Y, Z)$ on a rigid object in 3D to a point $p(x, y)$ in the retinal plane: u_x and u_y are the x and y components of the image velocity at a given image position (x, y), Z is 3D depth, and f is the camera focal length.

5.2.4 MOTION TRAJECTORY

Motion trajectory is a high-level feature associated with a moving region, defined as a spatiotemporal localization of one of its representative points (e.g., the centroid).

5.2.4.1 Salient Syntax and Semantics

The trajectory model is a first- or second-order piece-wise approximation along time, for each spatial dimension. The core of the description is a set of key points, representing the successive spatiotemporal positions of the described object (positions of one representative point of the object, such as its center of mass). They are defined by their coordinates in space (2D or 3D) and time. Additionally, interpolating parameters can be added to specify nonlinear interpolations between key points.

5.2.4.2 Highlights of Extraction

The extraction in this case relies on the availability of segmented spatiotemporal regions in the content. This would be either through the alpha channel of MPEG-4 or by carrying out a segmentation prior to computing the trajectory to mark the regions. Once the regions are segmented, the motion trajectory is defined as the locus of the center of mass of the region. Note that if the segmentation is already done, the descriptor is easy to extract.

5.2.5 PARAMETRIC MOTION

This descriptor addresses the motion of objects in video sequences, as well as global motion. If it is associated with a region, it can be used to specify the relationship between two or more feature point motion trajectories according to the underlying motion model. The descriptor characterizes the evolution of arbitrarily shaped regions over time in terms of a 2D geometric transform.

In parametric model, the descriptor expressions are as follows:

Translational model: $V_x(x, y) = a_1$, $V_y(x, y) = a_2$

Rotation/scaling model: $V_x(x, y) = a_1 + a_3x + a_4y$, $V_y(x, y) = a_2 - a_4x + a_3y$

Affine model: $V_x(x, y) = a_1 + a_3x + a_4y$, $V_y(x, y) = a_2 + a_5x + a_6y$

Planar perspective model: $V_x(x, y) = \frac{(a_1 + a_3x + a_4y)}{(1 + a_7x + a_8y)}$

Parabolic model:

$$V_x(x, y) = a_1 + a_3x + a_4y + a_7xy + a_9x^2 + a_{11}$$

$$V_y(x, y) = a_2 + a_5x + a_6y + a_8xy + a_{10} + a_{12}$$

where $V_x(x, y)$ and $V_y(x, y)$ represent the x and y displacement components of the pixel at coordinates (x, y). The descriptor should be associated with a spatiotemporal region. Therefore, along with the motion parameters, spatial and temporal information is provided.

5.2.5.1   Highlights of Extraction

This descriptor relies on computationally intense operations on the motion field to fit the various models. It is the most computationally complex of the MPEG-7 motion descriptors.

5.2.6   *MOTION ACTIVITY*

A human watching a video or animation sequence perceives it as being a slow sequence, a fast-paced sequence, an action sequence, and so on. The activity descriptor [88] captures this intuitive notion of "intensity of action" or "pace of action" in a video segment. Examples of high activity include scenes such as "goal scoring in a soccer match," "scoring in a basketball game," or "a high-speed car chase." On the other hand, scenes such as "news reader shot," "an interview scene," or "a still shot" are perceived as low action shots. Video content in general spans the gamut from high to low activity. This descriptor enables us to accurately express the activity of a given video sequence/shot and comprehensively covers the aforementioned gamut. It can be used in diverse applications such as content repurposing, surveillance, fast browsing, video abstracting, video editing, content-based querying, and so on.

To more efficiently enable the applications mentioned earlier, we need additional attributes of motion activity. Thus, our motion activity descriptor includes the following attributes.

5.2.6.1 Salient Semantics and Syntax

- **Intensity**. This attribute is expressed as a 3-bit integer lying in the range [70, 89]. The value of 1 specifies the lowest intensity, whereas the value of 5 specifies the highest intensity. Intensity is defined as the variance of motion vector magnitudes, first appropriately normalized by the frame resolution and then appropriately quantized as per frame resolution [88]. In other words, the thresholds for the 3-bit quantizer are normalized with respect to the frame resolution and rate.

- **Dominant direction**. This attribute expresses the dominant direction and can be expressed as an angle between 0 and 360 degrees.

- **Spatial distribution parameters:** N_{sr}, N_{mr}, N_{lr}. Short, medium, and long runs of zeros are elements of the motion activity descriptor that provide information about the number and size of active objects in the scene. Their values are extracted from the thresholded motion vector magnitude matrix, which has elements for each block indexed by (i, j). Each run is obtained by recording the length of zero runs in a raster scan order over this matrix. The thresholded motion vector magnitude matrix is given by the following: For each object or frame the "thresholded activity matrix" C_{mv}^{thresh} is defined as

$$C_{mv}^{thresh}(i, j) = \begin{cases} C_{mv}(i, j) & \text{if } C_{mv}(i, j) > C_{mv}^{avg} \\ 0 & \text{otherwise} \end{cases}$$

where the motion vector matrix and average motion vector magnitude for an M × N macroblock frame are defined as $C_{mv} = R(i, j)$ where $R(i, j) = \sqrt{x_{i,j}^2 + y_{i,j}^2}$ for interblocks and $R(i, j) = 0$ for intrablocks.

From the thresholded motion vector magnitude matrix, the zero run-lengths are classified into three categories: short, medium, and long, which are normalized with respect to the frame width (see "MPEG-7" [88] for details).

- **Spatial localization parameter(s)**. A 3-bit integer expressing the quantized percentage duration of activities of each location (i.e., division block).

- **Temporal parameters**. This is a histogram consisting of 5 bins. The histogram expresses the relative duration of different levels of activity in the sequence. Each value is the percentage of each quantized intensity-level segments duration compared to the whole sequence duration.

5.2.6.2   Highlights of Extraction

It is evident from the semantics that the motion activity descriptor relies on simple operations on the motion vector magnitudes. Most often these are MPEG-1/2/4 motion vectors. It lends itself therefore to compressed domain extraction and is also easy to compute since it does not rely on any preprocessing step such as segmentation. The intention of the descriptor is to capture the gross motion characteristics of the video segment. It is the least computationally complex of the MPEG-7 motion descriptors.

5.2.7   APPLICATIONS OF MOTION DESCRIPTORS

Possible applications of MPEG-7 motion descriptors are as follows:

(1) *Content-based querying and retrieval from video databases*. Since all the descriptors are compact, they would lend themselves well to this application.

(2) *Video browsing*. The motion activity descriptor can help find the most active parts of a soccer game for example. The camera motion descriptor would also be helpful, as would the motion trajectory descriptor.

(3) *Surveillance*. The basic aim in this case would be event detection in stored or live video. The motion activity descriptor is the easiest to apply, but the other descriptors could also be used to detect specific actions.

(4) *Video summarization*. We have shown that the intensity of the motion activity of a video sequence is in fact a direct indication of its summarizability [90]. In the next section, we describe an application that uses the summarizability notion to dynamically generate video summaries.

(5) *Video repurposing*. The motion descriptors can be used to control the presentation format of content—for example, by dropping more frames when the motion activity intensity is low and fewer when it is high, or by slowing down the replay frame rate when the camera

motion is a fast pan. While there are perhaps other possible applications for the MPEG-7 motion descriptors, some issues emerge from considering the previous applications:

- *Extraction complexity*. Low extraction complexity (i.e., fast generation of content description) is obviously desirable for all applications, but for applications such as consumer video browsing, it is essential. To this end, descriptors that lend themselves to computation with motion vectors extracted from MPEG-1/2/4 or other compressed video are the most desirable. The parametric motion descriptor and the motion trajectory descriptor require a dense motion field for accurate computation. The camera motion and motion activity descriptors can be successfully extracted from compressed domain motion vectors. Thus, they are the most suitable for video browsing applications. The wider scope of the motion activity descriptor makes it our first choice for incorporation into a video browsing system.

- *Inferring higher-level features from low-level features*. All four described descriptors are low-level descriptors. However, in the right context, they can provide hints to higher level features. For instance, the most interesting moments in a soccer game are often marked by high motion activity and camera motion. In a specific domain, a certain trajectory would immediately identify an event (e.g., in a game of tennis). Therefore, all four descriptors are highly applicable to systems that use lower-level features to infer higher-level features. Such systems typically have less stringent requirements on computational complexity since they are seldom in real time. Note that for a short video program, the motion activity and the camera motion descriptors already provide a simple way to infer higher-level features such as an interesting play. However, such inferences are still broad. In all likelihood, a domain-specific, knowledge-based inference superstructure would be required to detect semantic events using the MPEG-7 motion descriptors.

- *Content-based lower-level operations*. The MPEG-7 motion descriptors would be extremely useful in providing hints for lower-level operations such as transcoding from one bitrate to another, content enhancement, and content presentation.

5.2.8 VIDEO BROWSING SYSTEM BASED ON MOTION ACTIVITY

As mentioned earlier, the compactness and ease of extraction of the motion activity descriptor make it the best candidate for a video browsing system. In our application, we use the intensity and the spatial distribution attributes of the motion activity descriptor. Figure 5.3 illustrates finding the most active shots in a video program using the intensity of motion activity. Note that the results are a mixture of sports and other shots. Figure 5.4 illustrates the use of the spatial distribution attribute of the motion activity distributor. As the figure shows, the spatial distribution goes beyond the

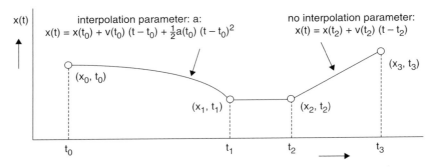

Figure 5.3   Motion trajectory.

Figure 5.4   Video browsing. Extracting the 10 most "active" video segments in a news program.

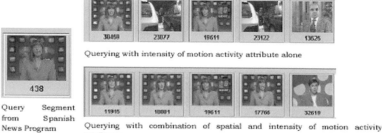

Figure 5.5 Illustration of video browsing with and without descriptors of spatial distribution of motion activity.

intensity of motion activity and helps match similar activities such as, in this case, talking. Figure 5.5 again illustrates the use of the spatial attribute in matching shots with similar activities within a news video program in the first row, since the query and the retrieval shots are all "head and shoulders" shots. In the next two rows, we show that we can use the motion activity descriptor as a first line of attack to prune the search space, before using a bulky (64-bin) color histogram to get the final matches in the subsequent stage. Note that the results are similar to those we get by using color alone in row 2. Note that the color histogram is also extracted in the compressed domain.

Figures 5.3, 5.4, and 5.5 illustrate the video indexing capabilities of the motion activity descriptor. We have shown that the intensity of motion activity is an indication of the summarizability of the video sequence [90]. This gives rise to a sampling approach to video summarization in which video sequences are subsampled heavily if the activity is low and not so heavily if the activity is high [88, 91]. We can thus vary the length of the summary of a shot by varying the subsampling rate. We illustrate this approach in Figure 5.6.

We illustrate our system in Figure 5.7. We uniformly sample the video program roughly every 1.5 seconds and extract features in the MPEG-1 compressed domain for each of those 1.5-second shots, as well as thumbnails. We do so to circumvent shot detection and to maintain one-pass feature extraction. We extract the four-element motion feature vector as mentioned earlier, as well as a color histogram taken from the direct component (DC) values of the first intracoded frame of the shot. The interface displays all the thumbnails, and each video shot can be played by clicking

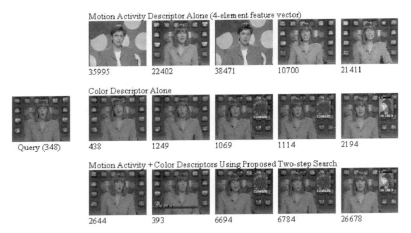

Figure 5.6   Video indexing using motion activity combined with color.

Figure 5.7   Illustration of adaptive subsampling approach to video summarization. The top row shows a uniform subsampling of the surveillance video, while the bottom row shows an adaptive subsampling of the surveillance video. Note that the adaptive approach captures the interesting events, while the uniform subsampling mostly captures the highway when it is empty. The measure of motion activity is the average motion vector magnitude in this case, which is similar to the MPEG-7 intensity of a motion activity descriptor in performance.

on the corresponding thumbnail. A query can be launched by right-clicking on the desired query thumbnail and choosing the desired combination of features. So far we have described the indexing or bottom-up traversal capabilities of the system. In addition, our system provides a dynamic summarization framework that relies on the intensity of motion activity. As Figure 5.8 shows, the user can specify any portion of the video program by entering the starting and stopping times and immediately receive a summary of the video sequence. The summary is displayed in the form of an array of key frames, but it can also be played as a concatenated video sequence. In this manner, the user can traverse the content from the top down until he or she gets the desired portion. The thumbnails received as a result of the summarization also lend themselves to the indexing described earlier. Thus, our system provides the user with extremely flexible traversal of the content using top-down (video summarization) or bottom-up (video indexing) traversal. Note that our system is Web based, and can thus serve clients of varying capabilities and bandwidths. We already demonstrated this system to the Japanese press and to the MPEG-7 Awareness event in Singapore in 2001.

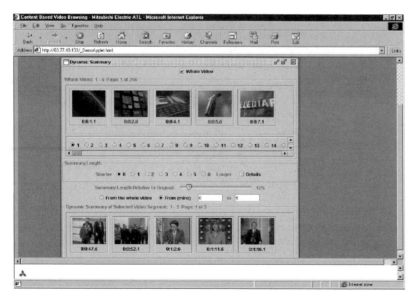

Figure 5.8 WWW-based interface for dynamic video summarization.

5.2.9 CONCLUSION

We presented an overview of the MPEG-7 motion descriptors and discussed their applications. We then motivated and presented a video browsing system based on motion activity. Our system is easy to use and computationally simple. In future work, we plan to incorporate the camera motion descriptor into the video browsing framework, as well as refine the existing summarization by adding audio features.

5.3 Indexing with Low-Level Features: Color

Color has been the feature of choice for video and image indexing, with the reason being perhaps that color descriptors are relatively easy to extract and lend themselves to simple interpretation. The most common color descriptor used in the literature is, of course, the color histogram, which directly captures the probability distribution of the color. We now provide an overview of selected MPEG-7 color descriptors.

- The *color space descriptor* enables the selection of the color space to be used.
- The *dominant color descriptor* enables specification of a small number of dominant color values as well as their statistical properties. Its purpose is to provide a simple description of the image.
- The *scalable color descriptor* is derived from a color histogram defined in the hue-saturation-value color scape. It uses a Haar transform encoding, thus allowing scalable representation as well as scalable computation for increasing or decreasing the accuracy of the matching and extraction. The commonly used color histogram finds its place in MPEG-7 through this descriptor. The *group of frames descriptor* is an extension of the scalable color descriptor to a group of frames in a video or a collection of pictures. This descriptor captures the overall color characteristics of the collection of still images or video frames.
- The *color structure descriptor* captures local color distributions in the image by scanning the image with a "structuring element," which is an 8×8 pixel square, and adapting the color structure histogram to the color characteristics of the 8×8 square. Note that if the square region were to consist of just one pixel, the descriptor would reduce to the typical histogram. The idea is therefore to preserve the local characteristics that are removed by the overall color histogram.

- The *color layout descriptor* expresses the spatial layout of the representative colors on a grid superimposed on a region or image, with the grid consisting of blocks of 8 × 8 discrete cosine transform coefficients. This is an extremely compact descriptor that enables fast browsing.

Many other color descriptors have been reported in the literature. However, the MPEG-7 collection of descriptors is a good indication of the challenges that have motivated the community to go beyond color histograms. On one hand, descriptors such as dominant color have sought to drastically reduce the size and complexity of the color description so as to enable rapid browsing. On the other hand, descriptors such as the color layout descriptor have sought to preserve the local color characteristics that are by definition lost in a global distribution such as the color histogram. Furthermore, with video sequences, it is often more important to capture the aggregate color characteristics of a collection of frames such as a shot, rather than the histogram of just one of the frames.

Color descriptors lend themselves well to computation in the compressed domain. The color layout descriptor, for instance, directly lies on the discrete cosine transform (DCT) coefficients that can be obtained from a still image without full decoding. Even with moving images, the DCT coefficients can be obtained with relatively low computation. The average values or DC values of each 8 × 8 block can then be used to construct an 8 × 8 reduced size image, which can then be used to extract color descriptors with much less computation than with the full image.

5.4 Indexing with Low-Level Features: Texture

Image texture is an important regional property that in certain applications, such as surveillance and medical imaging, can capture key semantic characteristics. Many common natural and artificial objects such as bodies of water, patches of grass, and carpets can be well characterized with their texture. An image can therefore also be considered a mosaic of homogeneous textures, thus enabling indexing using constituent textures. MPEG-7 has three kinds of texture descriptors that enable browsing and similarity retrieval.

- The *homogeneous texture descriptor* provides a quantitative representation using 62 numbers, consisting of the mean energy and the energy deviation from a set of frequency channels. This descriptor relies on a Gabor function decomposition of the texture in space frequency, since such Gabor function decomposition is thought to mimic early visual processing in the human cortex.

- The *texture browsing descriptor* captures the perceptual characteristics of a texture in terms of regularity, coarseness, and directionality. This descriptor is best for applications such as browsing, in which the user needs a quick approximation of a description.

- The *edge histogram descriptor* represents local-edge distribution in the image. The image is divided into 4 × 4 subimages and the local-edge distribution of each of the subimages is aggregated into a histogram.

The homogeneous texture-based descriptors (the first two in the preceding list) do implicitly rely on region-based processing such as sampling or segmentation. The edge histogram descriptor does have the ability to deal with nonhomogeneous textures. The homogeneous texture descriptor has been successful with aerial imagery and other surveillance applications. The texture browsing descriptor has also been used successfully for browsing textile patterns. The edge histogram descriptor has been tried successfully in query by sketch applications. From our point of view—namely, low complexity video summarization—texture descriptors represent a challenge in that they are expensive to compute but contain a wealth of information.

5.5   Indexing with Low-Level Features: Shape

Shape is a semantic concept that can have different meanings depending on the context. While most real-world objects are three dimensional, most imagery is two dimensional, and hence for such imagery we have to satisfy ourselves with descriptions of the 2D projections of the aforementioned 3D objects. We will cover the MPEG-7 region-based and contour-based 2D shape descriptors and then describe the MPEG-7 3D shape descriptor.

- The *region-based shape descriptor* interprets shape as the distribution of pixels within a 2D object or region. It belongs to the broad category of shape descriptors that rely on moments. Such descriptors are well known for examples in character recognition, which also capitalizes on the ability of these descriptors to describe complex objects consisting of multiple disconnected regions as well as simple objects with or without holes. A big advantage of these descriptions is that they are robust to segmentation errors as long as all subregions are used in the computation.

- The *contour-based shape descriptor* interprets shape as the local and global curvature properties of the contour of the object. It therefore is best used for objects whose shapes are efficiently described by their contours. Strengths of this descriptor include robustness to significant nonrigid deformations, compactness, and ability to describe shape scalably (i.e., to varying degrees of accuracy of approximation).

- The *3D shape descriptor* captures characteristic features of objects represented as discrete polygonal 3D meshes. It relies on the histogram of local geometric properties of the 3D surfaces of the object.

Indexing with shape descriptors relies on segmentation of the image. For certain applications that generate content in controlled conditions such as surveillance and search for easily segmented objects such as logos, segmentation can be carried out simply and reliably. While shape indexing is powerful, its reliance on accurate segmentation makes it difficult to use for consumer video browsing. As object-based coding catches on, that might change.

5.6   Indexing with Low-Level Features: Audio

To illustrate the importance of audio in video content, let us conduct a thought experiment. Let us first imagine watching the news without the sound and then listening to the news without the images. In this case, the audio by itself has the bulk of the semantic content. While this is obviously an extreme, it serves to emphasize that audio is a powerful complement to the visual content in a video sequence.

Audio indexing can be purely based on audio needs such as query-based humming, spoken-query retrieval, and consumer-level audio editing. Our interest in audio is from the point of view of considering video as a composite of visual and audio content and exploiting both the visual and the aural for meta-data extraction. The aforementioned applications provide a broad categorization of audio descriptors. Low-level audio descriptors, such as volume spectral descriptors, describe the signal characteristics. Higher-level descriptors, such as audio class descriptions, capture higher-level semantics. For instance, as we have described elsewhere in this book, we can recognize sound categories such as applause, cheering, and excited speech to extract sports highlights.

MPEG-7 provides both low-level and high-level descriptors.

(1) Low-level descriptors (LLD)

- *Basic* descriptors express instantaneous waveform and power values.
- *Basic spectral* descriptors express the log-frequency power spectrum and spectral features including spectral centroid, spectral spread, and spectral flatness.
- *Signal parameters* express the fundamental frequency of quasi-periodic signals and the harmonicity of signals.
- *Temporal timbral* descriptors consist of log attack time and the temporal centroid.
- *Spectral timbral* descriptors consist of spectral features in a linear-frequency space and spectral features specific to the harmonic portions of signals.
- *Spectral basis representations* consist of two features generally useful as projections into a low-dimensional space to aid compactness and recognition.

(2) High-level descriptors (HLD)

Note that the low-level audio descriptors target both common applications such as speech and music signal analysis as well as other applications such as generic sounds. MPEG-7 has three categories of high-level indexing tools for each of the aforementioned applications.

- *General sound recognition and indexing tools* are a collection for indexing and categorization of sound effects and sounds in general. Our work on summarization in particular has been strongly inspired by this category of tools. Support for automatic sound effect identification and indexing is included, as well as tools for specifying a taxonomy of sound classes and tools for specifying an ontology of sound recognizers.
- *Spoken content description tools* combine word and phone information so as to enable refinement of the speech recognition. This allows the retrieval to be robust to imperfect speech recognition since the low-level information is also retained.

- *Musical instrument timbre description tools* aim to describe perceptual features of instrument sounds. As mentioned earlier, there are numerous timbre-related features in the LLD set as well. While musical instrument sounds can be classified as harmonic, sustained, coherent sounds, nonharmonic sustained, coherent sounds, percussive, nonsustained sounds, and noncoherent, sustained sounds, MPEG-7 focuses on harmonic, coherent, sustained sounds and nonsustained, percussive sounds. The other two are not dealt with because of their presumed rarity.

In conclusion, audio features are a rich source of content semantics. Our evidence indicates that it is the higher-level audio classification that is best suited for video summarization applications. We will elaborate on these applications in other chapters. Audio analysis has two strong advantages in our view. First, audio data consume much less bandwidth compared to video data, and therefore audio analysis tends to be much less computationally complex. Second, audio classification enables much easier access to content semantics than does visual analysis. While audio classes such as applause and cheering are easily derived from low-level features, the visual counterparts such as, say, "walking people" are much harder to derive from low-level visual features. It is this advantage that has strongly motivated us to employ audio-only solutions for video summarization.

5.7 Indexing with User Feedback

The similarity measures for the low-level descriptors discussed in the previous sections are all based on objective distance and are not necessarily directly based on "semantic" or subjective distance. While the distance measure computation is carried out in the feature domain, the success of the similarity match is measured semantically since that is the only ground truth that can be created. In other words, there is a substantial gap between the low-level features and the content semantics. A way of bridging this gap has been to incorporate user feedback or "relevance feedback" into the content-based querying framework [92, 93]. The user specifies the relevance of the query results returned by the initial automatic retrieval, and the system refines its retrieval mechanism based on this relevance feedback. Many methods have been proposed in the literature. One of the well-known techniques is "active learning" [93], which consists of making the user focus on query results that lie close to the boundaries of the discriminated classes and are thus most in need of clarification. The result is that the user relevance feedback is provided exactly for those query responses that most need the human intervention, and thus the retrieval mechanism converges rapidly to optimal performance. Furthermore, as Tong and Chang point out, similarity metrics that treat the search space as a metric space oversimplify the problem, because there is evidence from cognitive psychology that similarity is both dynamic and nonmetric [93]. For instance, the attributes used to declare that images A and B are similar need not be the same as the attributes used to declare that the images A and C are similar, so that those two similarities need not result in any similarity between images B and C.

5.8   Indexing Using Concepts

Another approach to bridging the gap between low-level features and semantics has been to use machine learning to infer high-level concepts from low-level features. Notable examples are Naphade and Huang's work on deriving concepts such as "outdoors" and "indoors" [94] from low-level features of still images such as color and texture, Xie et al.'s work on modeling "play" and "break" segments of soccer video using hidden Markov models with motion activity and color features [6], and Casey's work on deriving high-level audio concepts such as "male voice," "shattering glass," and "trumpet" from low-level spectral features using hidden Markov models [95]. The aim of this work has been to abstract semantic concepts from the low-level features of multimedia content. Since the publication of the aforementioned papers, much related work has been published that has further broadened the work to cover other genres.

It is logical, therefore, to index video using concepts rather than low-level features [96]. The TREC (Chaisorn et al. [97]) video retrieval experiments have been exploring the same theme by assigning semantic retrieval tasks that have to be carried out without any preexisting semantic annotation. Naphade and Huang report that concept-based indexing actually beats indexing based on low-level features [94]. Their approach consists of deriving high-level concepts and then indexing content based on the percentage concept composition. For instance, a video may consist of 10% outdoors scenery, 80% walking people, and 10% news anchor. Thus, such a "concept histogram" can be used to compare different pieces of video content.

5.9 Discussion and Conclusions

The preceding sections show that video indexing lends itself to both low- and high-level feature-based approaches. The literature shows much work on combinations of features, unimodal as well as multimodal, to improve the results of indexing. The video indexing community has diligently examined the entire range of available low-level features and obtained interesting results. However, the evidence thus far indicates that video indexing needs to bridge the gap between low-level features and content semantics so as to make a significant impact, both scientifically and technologically. The next challenge is, therefore, to bridge the semantic gap and create compelling applications.

Chapter 6 | A Unified Framework for Video Summarization, Browsing, and Retrieval

As we have reviewed in the previous chapters, considerable progress has been made in each of the areas of video analysis, representation, browsing, and retrieval. However, so far the interaction among these components is still limited, and we still lack a unified framework to glue them together. This is especially crucial for video, given that the video medium is characteristically long and unstructured. This chapter explores the synergy between video browsing and retrieval.

6.1 Video Browsing

This section recapitulates key components described in Chapter 2. Among the many possible video representations, the scene-based representation is probably the most effective for meaningful video browsing. We have proposed a scene-based video table of contents (ToC) representation in Chapter 2. In this representation, a video clip is structured into the scene-group-shot-frame hierarchy (see Figure 1.2), which then serves as the basis for the ToC construction. This ToC frees the viewer from doing tedious fast forward and rewind operations and provides the viewer with nonlinear access to the video content. Figures 6.1 and 6.2 illustrate the browsing process, enabled by the video ToC. Figure 6.1 shows a condensed ToC for a video clip, as we normally have in a long book. By looking at the representative frames and text annotation, the viewer can determine which particular portion of the video clip he or she is interested in. Then, the viewer can further expand the ToC into more detailed levels, such as groups and shots. The expanded ToC is illustrated in Figure 6.2. Clicking on the Display button will display the specific portion that is of interest to the viewer, without viewing the entire video.

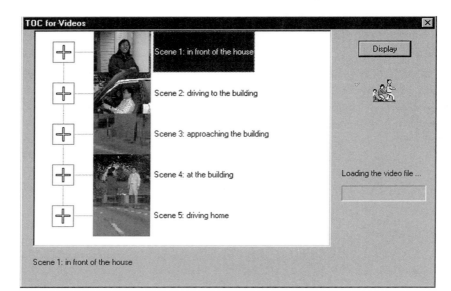

Figure 6.1   The condensed ToC.

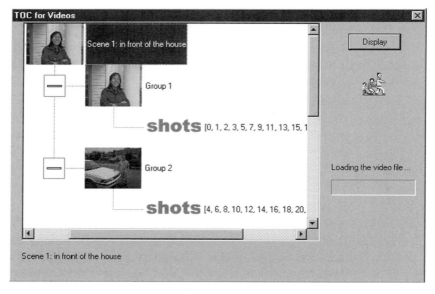

Figure 6.2   The expanded ToC.

6.2 Video Highlights Extraction

This section recapitulates the key components of video highlights extraction described in Chapter 3. We have shown the framework's effectiveness in three different sports: soccer, baseball, and golf. Our proposed framework can be summarized in Figure 6.3. Figure 6.3 has four major components. We describe them one by one in the following sections.

6.2.1 AUDIO MARKER DETECTION

Broadcast sports content usually includes audience reactions to the interesting moments of the games. Audience reaction classes, including applause, cheering, and the commentator's excited speech, can serve as audio markers. We have developed classification schemes that can achieve very high recognition accuracy on these key audio classes. Figure 6.4 shows our proposed unified audio marker detection framework for sports highlights extraction.

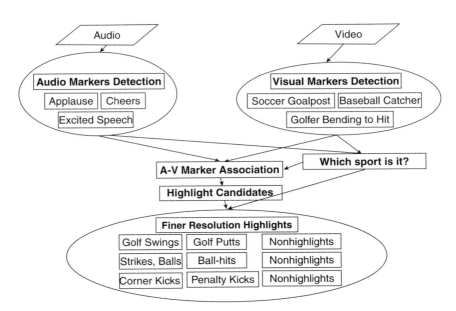

Figure 6.3 Proposed approach for sports highlights extraction.

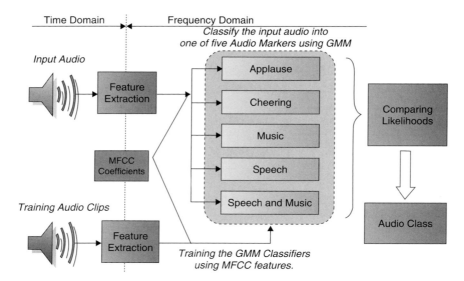

Figure 6.4   Audio markers for sports highlights extraction.

6.2.2   *VISUAL MARKER DETECTION*

As defined earlier, visual markers are key visual objects that indicate the interesting segments. Figure 6.5 shows examples of some visual markers for three different games. For baseball games, we want to detect the pattern in which the catcher squats waiting for the pitcher to pitch the ball; for golf

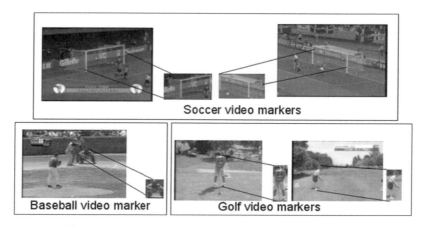

Figure 6.5   Examples of visual markers for different sports.

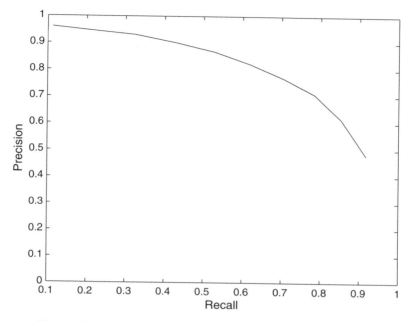

Figure 6.6 The precision-recall curve of baseball catcher detection.

games, we want to detect the players bending to hit the golf ball; for soccer, we want to detect the appearance of the goalpost. Correct detection of these key visual objects can eliminate most of the video content that is not in the vicinity of the interesting segments. For the goal of devising one general framework for all three sports, we use the following processing strategy: for the unknown sports content, we detect whether there are baseball catchers, or golfers bending to hit the ball, or soccer goalposts. The detection results can enable us to decide which sport (baseball, golf, or soccer) it is.

6.2.3 AUDIO-VISUAL MARKERS ASSOCIATION FOR HIGHLIGHTS CANDIDATES GENERATION

Ideally each visual marker can be associated with one and only one audio marker and vice versa. Thus, they make a pair of audio-visual markers indicating the occurrence of a highlight event in their vicinity. But since many pairs might be wrongly grouped due to false detections and misses, some postprocessing is needed to keep the error to a minimum. We perform the following for associating an audio marker with a video marker.

- If a contiguous sequence of visual markers overlaps with a contiguous sequence of audio markers by a large margin (e.g., the percentage of overlapping is greater than 50%), then we form a "highlight" segment spanning from the beginning of the visual marker sequence to the end of the audio-visual marker sequence.

- Otherwise, we associate a visual marker sequence with the nearest audio marker sequence that follows it if the duration between the two is less than a duration threshold (e.g., the average duration of a set of training "highlight" clips from baseball games).

6.2.4   FINER-RESOLUTION HIGHLIGHTS RECOGNITION AND VERIFICATION

Highlight candidates, delimited by the audio markers and visual markers, are quite diverse. For example, golf swings and putts share the same audio markers (audience applause and cheering) and visual markers (golfers bending to hit the ball). Both of these two kinds of golf highlight events can be found by the aforementioned audio-visual markers detection-based method. To support the task of retrieving finer events such as "golf swings only" or "golf putts only," we have developed techniques that model these events using low-level audio-visual features. Furthermore, some of these candidates might not be true highlights. We eliminate these false candidates using a finer-level highlight classification method. For example, for golf, we build models for golf swings, golf putts, and nonhighlights (neither swings nor putts) and use these models for highlights classification (swings or putts) and verification (highlights or nonhighlights).

As an example, let us look at a finer-level highlight classification for a baseball game using low-level color features. The diverse baseball highlight candidates found after the audio markers and visual markers negotiation step are further separated using the techniques described here. For baseball, there are two major categories of highlight candidates, the first being "balls or strikes" in which the batter does not hit the ball, the second being "ball-hits" in which the batter hits the ball into the field or audience. These two categories have different color patterns. In the first category, the camera is fixed at the pitch scene, so the variance of color distribution over time is low. In the second category, in contrast, the camera first shoots at the pitch scene, then it follows the ball to the field or the audience, so the variance of color distribution over time is higher.

6.3   Video Retrieval

This section recapitulates the key components of video retrieval described in Chapter 5. Video retrieval is concerned with how to return similar video clips (or scenes, shots, and frames) to a user given a video query. There are two major categories of existing work. One is to first extract key frames from the video data, then use image retrieval techniques to obtain the video data *indirectly*. Although easy to implement, it has the obvious problem of losing the temporal dimension. The other technique incorporates motion information (sometimes object tracking) into the retrieval process. Although this is a better technique, it requires the computationally expensive task of motion analysis. If object trajectories are to be supported, then this becomes more difficult.

Here we view video retrieval from a different angle. We seek to construct a video index to suit various users' needs. However, constructing a video index is far more complex than constructing an index for books. For books, the form of an index is fixed (e.g., key words). For videos, the viewer's interests may cover a wide range. Depending on his or her knowledge and profession, the viewer may be interested in semantic-level labels (buildings, cars, people), low-level visual features (color, texture, shape), or the camera motion effects (pan, zoom, rotation). In the system described here, we support the following three index categories:

- Visual index
- Semantic index
- Camera motion index

For scripted content, frame clusters are first constructed to provide indexing to support semantic-level and visual feature-based queries. For unscripted content, since audio marker detection and visual marker detection provide information about the content such as whether an image frame has a soccer goalpost or whether a segment of audio has the applause sound, we use images or audio segments with audio or video object detection as the visual index.

For scripted content, our clustering algorithm is described as follows:

(1) ***Feature extraction***. Color and texture features are extracted from each frame. The color feature is an 8×4 2D color histogram in HSV color space. The V component is not used because of its sensitivity to lighting conditions. The H component is quantized finer than the S component due to the psychological observation that the human

visual system is more sensitive to hue than to saturation. For texture features, the input image is fed into a wavelet filter bank and is then decomposed into de-correlated subbands. Each subband captures the feature of a given scale and orientation from the original image. Specifically, we decompose an image into three wavelet levels— thus, 10 subbands. For each subband, the standard deviation of the wavelet coefficients is extracted. The 10 standard deviations are used as the texture representation for the image [102].

(2) *Global clustering*. Based on the features extracted from each frame, the entire video clip is grouped into clusters. A detailed description of the clustering process can be found in Zhuang et al. [20]. Note that each cluster can contain frames from multiple shots, and each shot can contain multiple clusters. The cluster centroids are used as the visual index and can later be labeled as a semantic index (see Section 6.3.). This procedure is illustrated in Figure 6.7.

After this clustering process, all of the video clips are grouped into multiple clusters. Since color and texture features are used in the clustering process, all the entries in a given cluster are visually similar. Therefore, these clusters naturally support the visual queries.

For unscripted content, we group the video frames based on the key audio-visual marker detectors into clusters such as "frames with a catcher," "frames with a bat swing," and "clips with cheering." The video frames or audio clips are then used as visual index or audio index (see Figure 6.8).

To support semantic-level queries, semantic labels need to be provided for each cluster. There are two possible approaches. One is based on the hidden Markov model (HMM) and the other is an annotation-based approach. Since the former approach also needs training samples, both approaches are

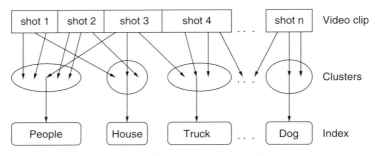

Figure 6.7   From video clip to cluster to index for scripted content.

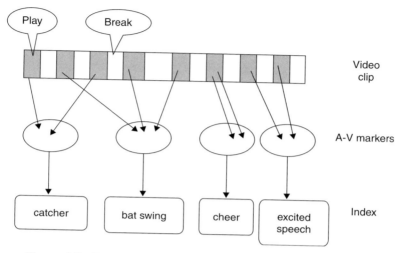

Figure 6.8   From video clip to cluster to index for unscripted content.

semiautomatic. To learn details of the first approach, readers are referred to Nephade et al. [17]. We will introduce the second approach here. Instead of attempting to attack the unsolved automatic image understanding problem, semiautomatic human assistance is used. We have built interactive tools to display each cluster centroid frame to a human user, who will label that frame. The label will then be *propagated* through the whole cluster. Since only the cluster centroid frame needs labeling, the interactive process is fast. For a 21,717-frame video clip (Movie1), about 20 minutes is needed. After this labeling process, the clusters can support both visual and semantic queries. The specific semantic labels for Movie1 are people, car, dog, tree, grass, road, building, house, and so on.

To support camera motion queries, we have developed techniques to detect camera motion in the MPEG-compressed domain [103]. The incoming MPEG stream does not need to be fully decompressed. The motion vectors in the bit stream form good estimates of camera motion effects. Hence, panning, zooming, and rotation effects can be efficiently detected [103].

6.4   A Unified Framework for Summarization, Browsing, and Retrieval

The previous three subsections described video browsing (using ToC generation and highlights extraction) and retrieval techniques separately.

In this section, we integrate them into a unified framework to enable a user to go back and forth between browsing and retrieval. Going from the index to the ToC or the highlights, a user can get the *context* where the indexed entity is located. Going from the ToC or the highlights to the index, a user can *pinpoint* specific queries. Figure 6.9 illustrates the unified framework.

An essential part of the unified framework is composed of the weighted links. The links can be established between index entities and scenes, groups, shots, and key frames in the ToC structure for scripted content and between index entities and finer-resolution highlights, highlight candidates, audio-visual markers, and plays/breaks.

For scripted content, as a first step, in this chapter we focus our attention on the links between index entities and shots. Shots are the building blocks of the ToC. Other links are generalizable from the shot link. To link shots and the *visual index*, we propose the following techniques. As we mentioned before, a cluster may contain frames from multiple shots. The frames from a particular shot form a subcluster. This subcluster's centroid is denoted as c_{sub} and the centroid of the whole cluster is denoted as c, as illustrated in Figure 6.10. Here c is a representative of the whole cluster

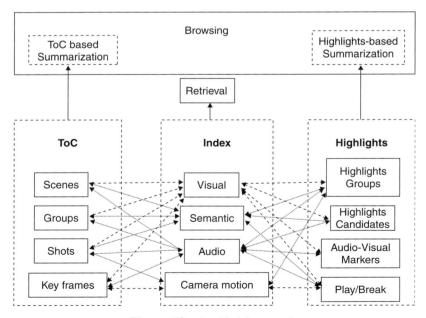

Figure 6.9   A unified framework.

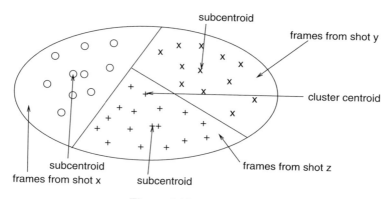

Figure 6.10   Subclusters.

(and thus the visual index), and c_{sub} is a representative of the frames from a given shot in this cluster. We define the similarity between the cluster centroid and subcluster centroid as the link weight between the index entity c and that shot:

$$w_v(i, j) = similarity(c_{sub}, c_j) \qquad (6.4.1)$$

where i and j are the indices for shots and clusters, respectively; and $w_v(i, j)$ denotes the link weight between shot i and visual index cluster c_j.

After defining the link weights between shots and the visual index, and labeling each cluster, we can next establish the link weights between shots and the *semantic index*. Note that multiple clusters may share the same semantic label. The link weight between a shot and a semantic index is defined as follows:

$$w_s(i, k) = max_j(w_v(i, j)) \qquad (6.4.2)$$

where k is the index for the semantic index entities; and j represents those clusters sharing the same semantic label k.

The link weight between shots and a *camera motion index* (e.g., panning) is defined as follows:

$$w_c(i, l) = \frac{n_i}{N_i} \qquad (6.4.3)$$

where l is the index for the camera operation index entities; n_i is the number of frames having that camera motion operation; and N_i is the number of frames in shot i.

For unscripted content, the link weight between plays/breaks and the *visual index* is defined as follows:

$$w_{p/b}(i, l) = \frac{m_i}{M_i} \qquad (6.4.4)$$

where l is the index for the audio-visual index entities (catcher, goal-post, cheering, etc.); m_i is the number of frames having that visual index detected; and M_i is the number of frames in play/break i.

We have carried out extensive tests using real-world video clips. The video streams are MPEG compressed, with the digitization rate equal to 30 frames/second. Table 6.1 summarizes example results over the video clip Movie1. The first two rows are an example of going from the semantic index (e.g., car) to the ToC (shots). The middle two rows are an example of going from the visual index to the ToC (shots). The last two rows are going from the camera operation index (panning) to the ToC (shots).

For unscripted content (a baseball game), we show an example of going from the visual index (e.g., a video frame with a catcher in Figure 6.11) to the highlights (segments with catcher(s)) in Figure 6.12. We show another example of going from one highlight segment to the visual index in Figure 6.13. Note that some frames without the catcher have also been chosen because there are audio markers (audience cheering) associated with them.

By just looking at each isolated index alone, a user usually cannot understand the context. By going from the index to the ToC or highlights (as in Table 6.1 and Figure 6.11), a user quickly learns when and under which

Table 6.1 **From the semantic, visual, camera index to the ToC.**

Shot ID	0	2	10	12	14	31	33
w_s	0.958	0.963	0.919	0.960	0.957	0.954	0.920
Shot ID	16	18	20	22	24	26	28
w_v	0.922	0.877	0.920	0.909	0.894	0.901	0.907
Shot ID	0	1	2	3	4	5	6
w_c	0.74	0.03	0.28	0.17	0.06	0.23	0.09

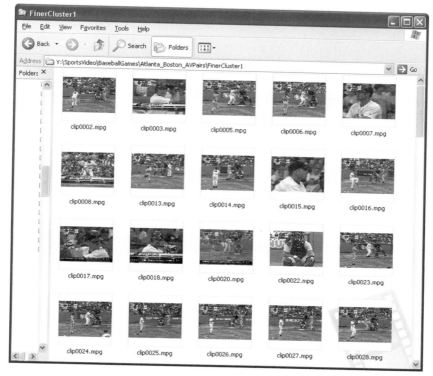

Figure 6.11 Interface for going from the semantic index to the highlights.

circumstances (e.g., within a particular scene) that index entity is happening. Table 6.1 summarizes how to go from the index to the ToC to find the *context*. We can also go from the ToC or the highlights to the index to *pinpoint* a specific index. Table 6.2 summarizes which index entities appeared in shot 33 of the video clip Movie1.

For a continuous and long medium such as video, a back-and-forth mechanism between summarization and retrieval is crucial. Video library users may have to see the summary of the video first before they know what to retrieve. On the other hand, after retrieving some video objects, the users will be better able to browse the video in the correct direction. We have carried out extensive subjective tests employing users from various disciplines. Their feedback indicates that this unified framework greatly facilitated their access to video content—in home entertainment, sports, and educational applications.

Figure 6.12   An image used as a visual index (catcher detected).

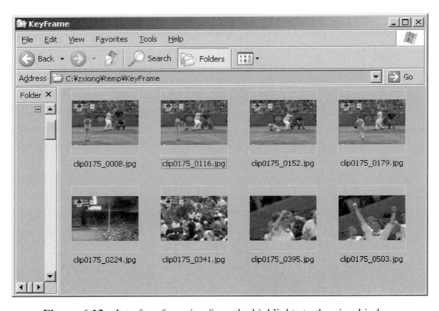

Figure 6.13   Interface for going from the highlights to the visual index.

Table 6.2   **From the ToC (shots) to the index.**

Index	Fence	Mailbox	Human hand	Mirror	Steering wheel
Weight	0.927	0.959	0.918	0.959	0.916

6.5   Conclusions and Promising Research Directions

In this chapter, we covered the following areas:

- Reviewed and discussed recent research progress in multimodal (audio-visual) analysis, representation, summarization, browsing, and retrieval;
- Introduced the video ToC, the highlights, and the index and presented techniques for constructing them;
- Proposed a unified framework for video summarization, browsing, and retrieval and proposed techniques for establishing the link weights between the ToC, the highlights, and the index.

We should be aware that video is not just an audio-visual medium. It contains additional text information and is thus true multimedia. We need to further extend our investigation to the integration of closed-captioning into our algorithm to enhance the construction of ToCs, highlights, indexes, and link weights.

Chapter 7 | Applications

7.1 Introduction

One of the reasons why video summarization, browsing, and retrieval has attracted so many research efforts and sustained development since the mid-1980s is its great potential in government and commercial applications. Since the mid-1980s, both the number of video summarization, browsing, and retrieval systems and the number of commercial enterprises have greatly increased due to the emergence of many new application areas; further improvement of the video summarization, browsing, and retrieval technologies; and increased affordability of the systems.

One application that stands out of many others is the summarization, browsing, and retrieval of broadcast sports video. Broadcast sports content usually includes audience reactions to the development of the games. These reactions, especially audio reactions such as applause or cheering, can be highly correlated with the interesting segments of the games. A reasonable solution based on an audio marker classification approach has been adopted by Mitsubishi Electric in its DVD and HDD recorder products.

This chapter reviews many of the video summarization, browsing, and retrieval applications that have already used these technologies. These applications are grouped into three categories, as shown in Table 7.1. Although we try to cover as many categories as possible, these three categories are not meant to be exclusive of each other or exhaustive. For each category, some of the exemplar applications are also listed. Sections 7.2 to 7.4 review these three categories. Section 7.5 reviews some of the limitations of the video summarization, browsing, and retrieval technologies. Section 7.6 presents our concluding remarks.

Table 7.1   **Application categories.**

Categories	*Exemplar Application Scenarios*
Consumer video	Sports, news, sitcom, movies, digital video recorder, DVD, home network, enhanced Internet multimedia service
Image/Video databases management	Video search engine, digital video library, object indexing and retrieval, automatic object labeling, object classification
Surveillance	Traffic surveillance, elevator surveillance, airports surveillance, nuclear plant surveillance, park surveillance, neighborhood watch, power grid surveillance, closed-circuit television control, portal control

7.2   Consumer Video Applications

In the recent years, there has been an increase in the storage and computational capacity of consumer electronic devices such as personal video recorders (PVR). Therefore, by enabling these devices with content analysis technologies, we can let the end user browse the recorded content in efficient ways. For instance, by detecting "interesting" highlight segments from a recorded sports program automatically, we enable the end user to view the interesting parts quickly. For scripted consumer video content such as situation comedies and news, by detecting scene changes and story boundaries, we can enable the end user to browse through the content efficiently with a table-of-contents (ToC)–based structure. We will discuss some example application scenarios here. The application scenarios are all instantiations of the same Audio-Visual Marker detection framework proposed in this book. They only differ where several parts of the architecture for the proposed analysis one realized.

- *Summarization-enabled PVR.* Figure 7.1 shows the architecture of a summarization-enabled digital video recorder. Here the analysis is performed while recording the video using compressed domain feature extraction and an audio classification framework.

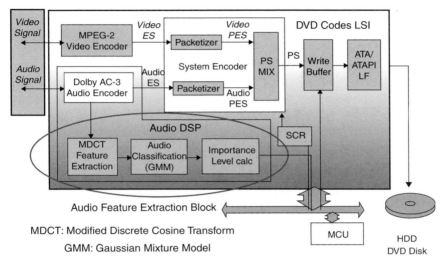

Broadcast
video

Figure 7.1   Architecture of a summarization-enabled DVD recorder.

The video and audio signals from a broadcast video are encoded using MPEG-2 and AC-3, packetized, and stored onto a disk such as hard disk drive (HDD), DVD, or Blu-ray medium via buffer. The audio classification block in the audio digital signal processing (DSP) classifies each audio segment as one of several classes (e.g., applause, cheering, excited speech, normal speech, music, etc.) using low-complexity Gaussian mixture models (GMM). The importance calculation block calculates the percentage of the significant audio class in a time window of an audio classification data stream to get the importance level for each audio segment. For instance, in sports the significant audio class turns out to be the audience and commentator reaction classes such as applause, cheering, and excited speech. From the user-interface point of view, a user can set a threshold on the importance for a desired summary length. The portions with importance greater than the chosen threshold are identified as the highlights. So skipping to the start position of the highlight scene manually and skipping and playing back only the highlights are functions that are supported. These unique functions give the user a powerful and useful way to browse large volumes of content.

Figure 7.2   Prototype of a summarization-enabled DVD recorder.

In the Japanese market, Mitsubishi Electric has released one such summarization enabled DVD recorder with automatic sports highlights extraction capability based on an audio marker detection framework. Figure 7.2 shows the prototype of this product. The product, with a storage capacity of 250GB, lets the user record broadcast sports video. The analysis for audio marker detection is performed while recording the content onto the local storage. Within a few seconds after the recording, the extracted highlights are available for the end-user to browse the content. The user can generate scalable summaries of varying lengths based on the importance level plot shown on the screen. The product has been able to reliably detect highlight scenes from contents of baseball, soccer, horse racing, and sumo wrestling. Hitachi also has released a Media-PC based sports highlights extraction system based on audio-visual analysis. These two products are good examples of successful commercialization of content-analysis technologies for video browsing in consumer electronic devices. It remains to be seen how these products will be received by the end-users.

- *Multimedia home network with a summarization server*. In recent years, wireless access technologies used in homes, such as 802.11 and the ultra wide band (UWB), have been increasing the usable

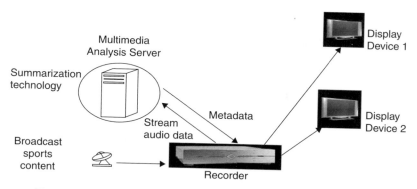

Figure 7.3   Architecture of a summarization-enabled home network.

bandwidth at home and thus can enable multimedia home networking applications. Figure 7.3 shows the architecture of a summarization-enabled multimedia home network architecture. Here the content analysis is performed on a set-top box or a multimedia home server after recording the program onto an HDD. Then, the content bit stream to be analyzed is streamed to the analysis server that has the content analysis technologies. The results of summarization (meta-data for the content) are sent back to one or more display devices at home.

- *Enhanced multimedia service over the Internet.* In the previous case, we considered the content analysis server to be a part of the home network. However, if the content provider can employ one of these servers to create meta-data for summaries, one can imagine delivering only the interesting parts of the content to the end user. For instance, using mobile phones with video capability, one can request a summary to the network service provider for a recent highlight scene of a live game. The request gets forwarded to the content analysis server, and the selected highlight segment can be transcoded to adapt to the display size of the mobile phone and delivered. Figure 7.4 shows the architecture of a network supporting the creation and the distribution of summarized content.

7.2.1   CHALLENGES FOR CONSUMER VIDEO BROWSING APPLICATIONS

As pointed out in the previous section, there are two products in the consumer electronic market for sports highlights extraction based on content

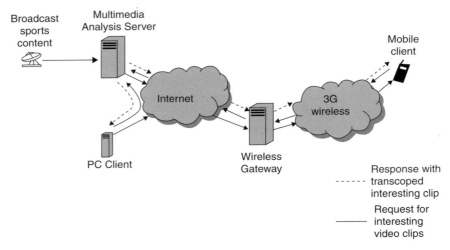

Figure 7.4 Architecture for summarization-enhanced multimedia services over the Internet.

analysis technologies. Even though the sports content is unconstrained, the presence of spontaneous audience reaction to interesting events and the ability to reliably detect them has helped bridge the gap between semantics and low-level feature analysis, thereby making it feasible to realize the application in these products. However, for other genres, such as movie and drama content, the gap still remains. For example, there are no laugh tracks in a comedy's audio track for us to detect comical moments in the movie. Therefore, one has to come up with intelligent audio-visual markers to support browsing applications for other consumer video genres. Also, there is a gap between the computational demands of the state-of-the art video content analysis technology and the computational capacity of typical embedded platforms in consumer electronic devices. The challenge is to devise computationally simple algorithms that are realizable on current consumer electronic product platforms.

7.3   Image/Video Database Management

In the early 1990s, because of the emergence of large-scale image collections, difficulties faced by the text-based image retrieval became more and more acute [104]. Content-based image/video retrieval tries to solve the difficulties faced by text-based image retrieval. Instead of being manually annotated by text-based key words, images would be indexed by their own visual content, such as color or texture. Feature (content) is the basis of content-based image retrieval, which usually uses the general features, such as color and texture. However, these general features have their own limitations. Recently researchers have tried to combine feature content with other image analysis technologies, such as "marker object" (e.g., face, car, trademark logo) detection and recognition to improve the retrieval accuracy. Human faces are shown in news, sports, films, home video, and other multimedia content, so faces can be good markers for the purpose of multimedia management. Indexing these multimedia contents by face detection, face tracking, face recognition, and face change detection is important for generating segments of coherent video content for video browsing, skimming, and summarization. It is also important to retrieve and index objects in object-only databases, for example, searching in mug books of criminals. In the following we give more exemplar applications that are based on "marker object" detection and recognition.

(1) A personal digital photo album has many images that either have human faces or have no human faces. Deciding whether an image has faces or not can be a preprocessing step to limit the range of search space for a given image query. The FotoFile [105] is one of the systems that try to support this functionality to make the managing of personal photo albums easier. The system also blends human and automatic annotation methods. The system prototypes a number of techniques, which make it easier for consumers to manually annotate content and to fit the annotation task more naturally into the flow of activities that consumers find enjoyable. The use of automated feature extraction tools enables FotoFile to generate some of the annotations that would otherwise have to be entered manually. It also provides novel capabilities for content creation and organization. When given photos that contain faces of new people, the face recognition system attempts to match the identity of the face. The user either corrects or confirms the choice; the system then can more accurately match faces to their correct identities in subsequent photos. Once a face is matched to a name, that name will be assigned as an annotation to all subsequently seen photos that contain faces that match the original. To handle the false positives and false negatives of the face recognition system, a user must

confirm face matches before the annotations associated with these faces are validated.

(2) One integrated multimedia management system is the Infomedia project at Carnegie Mellon University [106]. This project aims to create an information digital video library to enhance learning for people at all ages. Thousands of hours of video content are indexed and archived for search and retrieval by users via desktop computers through computer networks. One of its indexing schemes is face detection, developed by Rowley et al. [107]. The detected human faces and text are used as a basis for significance during the creation of video segments. A small number of face images can be extracted to represent the entire segment of video containing an individual for video summarization purposes. The system supports queries like "find video with talking heads" supported by face detection, "find interviews by Tim Russert" supported by face detection, and video text recognition.

(3) One image retrieval system, "Image-Seeker" by LTU Technologies, has been chosen by the French Patent Office (INPI) to facilitate the search for existing designs for preserving intellectual property. The patent office uses the retrieval system to visually compare new patent applications to the 200,000 existing patented designs and trademark logos.

From the aforementioned exemplar retrieval applications, it is clear that successful image retrieval applications have been made possible by narrowing the scope of the application. For instance, the "image seeker" by LTU Technologies works with an image database of trademark logos. Due to the semantic gap between low-level features and high-level concepts, designing an image retrieval system that works for a broader set of applications is still a challenge.

7.4 Surveillance

Today more than ever, security is a primary concern for airports, airline personnel, and passengers. Airport security systems that use object recognition technology have been seen in many airports around the globe. Figure 7.5 shows the diagram of an exemplar airport security system employing object recognition technology. Although it is possible to control lighting conditions and object orientation in some security applications—for example,

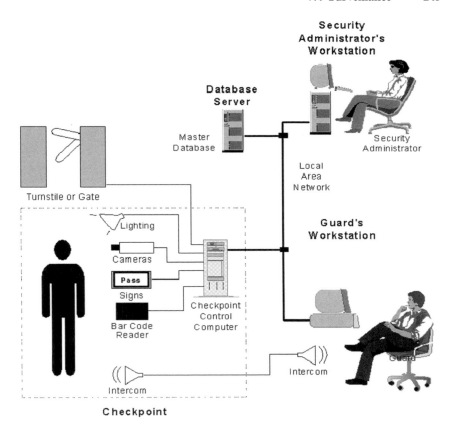

Figure 7.5 An exemplar airport security system.

using a single pedestrian lane with controlled lighting—one of the greatest challenges for object recognition in public places is the large number of objects that need to be examined, resulting in a high false alarm rate. Overall the performance of most of the recognition systems has not met the strict high recognition with low false alarm requirement. The user satisfaction level for this area of application is low.

Some of the exemplar systems in airports and stadiums are listed as follows:

(1) In October 2001, Fresno Yosemite International (FYI) Airport in California began to use Viisage's face recognition technology to deploy face recognition for airport security purposes. The system

is designed to alert FYI's Airport Public Safety Officers whenever an individual matching the appearance of a known terrorist suspect enters the airport's security checkpoint. Anyone recognized by the system would undergo further investigation by public safety officers.

(2) At Sydney Airport, Australian authorities are trying out a computerized face recognition system called SmartFace by Visionics, which uses cameras that have wide-angle lenses. The cameras sweep across the faces of all arriving passengers and send the images to a computer, which matches the faces with pictures of wanted people stored in its memory. If the computer matches the face with that of a known person, the operator of the surveillance system receives a silent alarm and alerts officers that the person should be questioned. The technology is also used at Iceland's Keflavik Airport to seek out known terrorists.

(3) In Oakland Airport, California, a face recognition system by Imagis Technologies of Vancouver, British Columbia, Canada, is used in interrogation rooms behind the scenes to match suspects brought in for questioning to a database of wanted criminals' pictures.

(4) Malaysia's 16 airports use a FaceIt-based security system to enhance passenger and baggage security. A lipstick-sized camera at the baggage check-in desk captures live video of the passenger and embeds the data on a smart-card chip. The chip is embedded on the boarding pass and on the luggage claim checks. The system ensures that only passengers who have checked their luggage can enter the departure lounge and board the aircraft, and that only luggage from boarding passengers is loaded into the cargo area. During the boarding process, the system automatically checks a real-time image of a passenger's face against that on the boarding pass smart chip. No luggage is loaded unless there is a match.

(5) Visage's faceFINDER equipment and software were used to scan the stadium audience at the 2001 Super Bowl at Raymond James Stadium in Tampa, Florida, in search of criminals. Video cameras set up at the entrance turnstiles scanned everyone entering the stadium. These cameras were tied to a temporary law-enforcement command center that digitized faces and compared them to photographic lists of known malefactors. The system is also used by South Wales Police in Australia to spot soccer hooligans who are banned from attending matches.

7.5   Challenges of Current Applications

Although video summarization, browsing, and retrieval technology has great potential in the applications just discussed, currently the scope of the applications is still limited. At least four challenges need to be addressed in order to gain large-scale application.

(1) The technology is still not robust, especially in unconstrained environments. "Marker object" recognition accuracy is not acceptable, especially for large-scale applications. Lighting changes, pose changes, and time differences between the probe image and the gallery image(s) further degrade the performance. The difficulty in detecting golfers as discussed in Chapter 3, is due to these constrained conditions.

(2) The deployment of surveillance systems causes some people to express concern about possible privacy violation. For example, the American Civil Liberties Union (ACLU) opposes uses of face recognition software in airports due to ineffectiveness and privacy concerns [108].

7.6   Conclusions

We have reviewed many video summarization, browsing, and retrieval systems in different application scenarios. We also pointed out the challenges of the current video summarization, browsing, and retrieval technology. The technology has evolved from laboratory research to many small-, medium-, or large-scale commercial deployments. At present, it is successful in small- or medium-scale applications such as highlight generation for a particular sport such as baseball and DVD ToC generation; it still faces great technical challenges for large-scale deployments such as airport security and general surveillance. With more research collaborations worldwide between universities and industrial researchers, the technology will become more reliable and robust.

Another direction for improving accuracy lies in a combination of multiple modalities and the "human in the loop" concept. For example, for security purpose at airports, these systems can work together with X-ray luggage scanners, metal detectors, and chemical trace detectors at security checkpoints.

Chapter 8 | Conclusions

The goal of this book is to develop novel techniques for constructing the video ToC, video highlights, and the video index as well as how to integrate them into a unified framework. To achieve these goals, we have introduced a hierarchical representation that includes key frames, shots, groups, and scenes for scripted video and another hierarchical representation that includes play/break, audio-visual markers, highlight candidates, and highlight groups for unscripted video. We also have presented a state-of-the-art content-adaptive representation framework on unsupervised analysis for summarization and browsing. We have shown how this framework can support supervised analysis as well. We have reviewed different video indexing techniques that are based on color, texture, shape, spatial layout, and motion activity. We then presented a unique and unified framework for video summarization, browsing, and retrieval to support going back and forth between the video ToC and the video index for scripted video, and between video highlights and video index for unscripted video. Furthermore, we have reviewed many of the video summarization, browsing, and retrieval applications that have already used these technologies.

For sports highlights extraction, we have shown that detection of audience reaction using audio markers is a reasonable solution. However, we should be aware that a change of genre, e.g., from sports to sitcoms, will make the problem suddenly more complex, partly because of new audio classes, e.g., laugher. How to minimize this sudden increase of difficulty is going to be one of the imminent tasks. In addition, easy-to-use interfaces will also play an important role in bringing applications in video summarization, browsing, and retrieval closer to the consumers. Although this issue has been briefly touched upon in Chapter 7, more research is needed in the near future.

Although for video highlight extraction we have shown some success in detecting objects by using some of the domain constraints, we should be aware that, in general, visual objects in video are seldom constrained in pose, lighting conditions, or occlusion. Furthermore, it is thought provoking to ask the following question: Do the sports video editors at television stations generate summaries of sports highlights using the approach we have proposed in this book? The answer is no. They solve the problem by using a cognitive approach that involves semantics and domain knowledge. For example, the domain knowledge they might use for a soccer goal event is that a soccer ball goes to the net. It involves object detection and the spatiotemporal relationships among objects. Our approach to sports video analysis has only addressed object detection, not the other parts (e.g., the representation of the domain knowledge and spatiotemporal relationship among objects).

Domain knowledge and the spatiotemporal relationship can be represented by state machines, predicate calculus, grammars and rules, and so on [109]. For example, a tree structure called "game tree" is used to represent the rules of chess [110], context-free grammars are used to represent the domains of some speech understanding problems [64], and Bayesian networks are used to model audio-visual signals [111]. Highlights of a chess game can be found by locating those branches of the tree with high "cost" values (or, "reward" values) in the Minimax or the Alpha-Beta algorithm [109, 110]. Identification of some targeted speech segments can be found by locating the corresponding decoded state sequence via the dynamic programming algorithm [64].

Videos can potentially use top-down representations similar to context-free grammars or Bayesian networks. Together with bottom-up signal processing, top-down knowledge representations will be promising for audio-visual scene analysis. Greater research efforts are needed in this important area.

Furthermore, audio object detection in mixed audio signals also remains challenging. One possibility is to use generative probabilistic models (GPMs) to separate the audio mixture in order to help audio analysis from a "cleaned-up" signal. How to incorporate that technology into the current framework also needs further research.

In this book we have reviewed some of the important issues in video summarization, browsing, and retrieval. Now we step back and try to ascertain: What are the most challenging research problems facing us in video summarization, browsing, and retrieval? Here is our top 10 list:

1) Bridging the Semantic Gap

Perhaps the most desirable mode of image/video retrieval is still by keywords or phrases. However, manually annotating the images and video data is extremely tedious. To do annotation automatically or semi-automatically, we need to bridge the "semantic gap," i.e., to find algorithms that will infer high level semantic concepts (sites, objects, events) from low level image/video features that can be easily extracted from the data (color, texture, shape and structure, layout; motion; audio pitch, energy, etc.). One sub-problem is Audio Scene Analysis. Researchers have worked on Visual Scene Analysis (Computer Vision) for many years, but Audio Scene Analysis is still in its infancy and is an under-explored field. Another sub-problem is multimodal fusion, particularly how to combine visual and audio cues to bridge the semantic gap in video.

2) How to Best Combine Human Intelligence and Machine Intelligence

One advantage of information retrieval is that in most scenarios there is a human (or humans) in the loop. One prominent example of human computer interaction is Relevance Feedback. Humans are good at processing small amounts of data in a highly sophisticated way, while machines are good at processing large amounts of data in a much less sophisticated way. The key is in combining the strengths of humans and machines.

3) New Query Paradigms

For image/video retrieval, people have tried query by keywords, similarity, sketching an object, sketching a trajectory, painting a rough image, etc. Each of these query paradigms is insufficient by itself. The challenge lies in coming up with new query paradigms that can work with multiple modes and also with ill-formed queries.

4) Multimedia Data Mining

Early results from Chapter 4 show that it is possible to discover anomolous pattern from multimedia. The challenge is to extend to large volumes of content and to discover a wider range of patterns.

5) How to Use Unlabeled Data?

We can consider Relevance Feedback as a two-category classification problem (relevant or irrelevant). However, the number of training samples is very small. Can we use the large number of unlabeled samples in the database to help? Also, how about active learning (to choose the best samples to return to the user to get the most information about the user's intention through feedback)?

Another problem related to image/video data annotation is Label Propagation. Can we label a small set of data and let the labels propagate to the unlabeled samples?

6) Incremental Learning

In most applications, we keep adding new data to the database. We should be able to change the parameters of the retrieval algorithms incrementally, rather than needing to start from scratch every time we have new data.

7) Using Virtual Reality Visualization to Help

Can we use 3D audio-visual visualization techniques to help a user navigate through the data space to browse and to retrieve?

8) Structuring Very Large Databases

Researchers in audio-visual scene analysis and those in Databases and Information Retrieval should really collaborate *closely* to find good ways of structuring very large multimedia databases for efficient retrieval and search.

9) Performance Evaluation

How do we compare the performances of different retrieval algorithms? Although TRECVID (video track of Text Retrieval Conference) has encouraged research in information retrieval by providing a large test collection and uniform scoring procedures for algorithms on this dataset, there is still the challenge of identifying metrics that measure human satisfaction of the retrieval and summarization results.

10) What Are the Killer Applications of Multimedia Retrieval?

Few real applications of multimedia retrieval have been accepted by the general public so far. Is sports highlight extraction, medical database retrieval, or a web multimedia search engine going to be the next killer application? It remains to be seen. With no clear answer to this question, it is still a challenge to do research that is appropriate for real applications.

Bibliography

1. H. Zhang, A. Kankanhalli, and S. W. Smoliar, "Automatic partitioning of full-motion video," *ACM Multimedia System Journal*, vol. 1, no. 1, pp. 1–12, 1993.
2. R. M. Bolle, B.-L. Yeo, and M. M. Yeung, "Video query: Beyond the keywords," Technical report, IBM Research, Oct. 1996.
3. D. Zhong, H. Zhang, and S.-F. Chang, "Clustering methods for video browsing and annotation," Technical report, Columbia University, 1997.
4. Y. Rui, T. S. Huang, and S. Mehrotra, "Exploring video structures beyond the shots," in *Proc. of IEEE Conf. Multimedia Computing and Systems*, 1998.
5. H. Zhang, S. W. Smoliar, and J. J. Wu, "Content-based video browsing tools," in *Proc. IS&T/SPIE Conf. on Multimedia Computing and Networking*, 1995.
6. L. Xie, P. Xu, S. F. Chang, A. Divakaran, and H. Sun, "Structure analysis of soccer video with domain knowledge and hidden Markov models," *Pattern Recognition Letters*, vol. 25, no. 7, pp. 767–775, May 2004.
7. A. Hampapur, R. Jain, and T. Weymouth, "Digital video segmentation," in *Proc. ACM Conf. on Multimedia*, 1994.
8. R. Kasturi and R. Jain, "Dynamic vision," in *Proc. of Computer Vision: Principles*, R. Kasturi and R. Jain, eds. 1991, IEEE Computer Society Press.
9. F. Arman, A. Hsu, and M.-Y. Chiu, "Feature management for large video databases," in *Proc. SPIE Storage and Retrieval for Image and Video Databases*, 1993.
10. J. Meng, Y. Juan, and S.-F. Chang, "Scene change detection in an MPEG compressed video sequence," in *Proc. SPIE Symposium on Electronic Imaging: Science & Technology-Digital Video Compression: Algorithms and Technologies*, vol. 241, pp. 14–25, 1995.
11. B.-L. Yeo, *Efficient processing of compressed images and video*, Ph.D. dissertation, Princeton University, 1996.
12. R. Zabih, J. Miller, and K. Mai, "A feature-based algorithm for detecting and classifying scene breaks," in *Proc. ACM Conf. on Multimedia*, 1995.
13. A. Nagasaka and Y. Tanaka, "Automatic video indexing and full-video search for object appearances," in *Proc. Visual Database Systems II*, 1992, pp. 113–127, Elsevier Science Publishers, Budapest, Hungary.
14. D. Swanberg, C.-F. Shu, and R. Jain, "Knowledge guided parsing in video databases," in *Proc. SPIE Storage and Retrieval for Image and Video Databases*, 1993.
15. H. Zhang and S. W. Smoliar, "Developing power tools for video indexing and retrieval," in *Proc. SPIE Storage and Retrieval for Image and Video Databases*, 1994.

16. H. Zhang, C. Y. Low, S. W. Smoliar, and D. Zhong, "Video parsing, retrieval and browsing: An integrated and content-based solution," in *Proc. ACM Conf. on Multimedia*, 1995.

17. M. R. Naphade, R. Mehrotra, A. M. Ferman, T. S. Huang, and A. M. Tekalp, "A high performance algorithm for shot boundary detection using multiple cues," in *Proc. IEEE Int. Conf. on Image Proc.*, Oct. 1998.

18. J. S. Boreczky and L. A. Rowe, "Comparison of video shot boundary detection techniques," in *Proc. SPIE Storage and Retrieval for Image and Video Databases*, 1996.

19. R. M. Ford, C. Robson, D. Temple, and M. Gerlach, "Metrics for scene change detection in digital video sequences," in *Proc. IEEE Conf. on Multimedia Comput. and Sys.*, 1997.

20. Y. Zhuang, Y. Rui, T. S. Huang, and S. Mehrotra, "Adaptive key frame extraction using unsupervised clustering," in *Proc. IEEE Int. Conf. on Image Proc.*, 1998.

21. P. O. Gresle and T. S. Huang, "Gisting of video documents: A key frames selection algorithm using relative activity measure," in *Proc. the 2nd Int. Conf. on Visual Information Systems*, 1997.

22. W. Wolf, "Key frame selection by motion analysis," in *Proc. IEEE Int. Conf. Acoust., Speech, and Signal Proc.*, 1996.

23. A. Divakaran, K. A. Peker, R. Radhakrishnan, Z. Xiong, and R. Cabasson, "Video summarization using MPEG-7 motion activity and audio descriptors," *Video Mining*, January 2003, A. Rosenfeld, D. Doermann and D. DeMenthon, eds., Kluwer Academic Publishers, Boston, MA.

24. L. Xie, S. F. Chang, A. Divakaran, and H. Sun, "Structure analysis of soccer video with hidden Markov models," in *Proc. Int'l Conf. on Acoustic, Speech, and Signal Processing*, vol. 4, pp. 4096–4099, May 2002.

25. C. Wren, A. Azarbayejani, T. Darrell, and A. Pentland, "Pfinder: Real-time tracking of the human body," *IEEE Transactions on Pattern Analysis and Machine Intelligence*, vol. 19, no. 7, pp. 780–785, July 1997.

26. Z. Xiong, R. Radhakrishnan, and A. Divakaran, "Effective and efficient sports highlights extraction using the minimum description length criterion in selecting GMM structures," in *Proc. Int'l Conf. on Multimedia and Expo*, June 2004.

27. Y. Rui, A. Gupta, and A. Acero, "Automatically extracting highlights for TV baseball programs," in *Proc. Eighth ACM Int'l Conf. on Multimedia*, pp. 105–115, 2000.

28. S. Nepal, U. Srinivasan, and G. Reynolds, "Automatic detection of 'goal' segments in basketball videos," in *Proc. ACM Conf. on Multimedia*, pp. 261–269, 2001.

29. Y.-L. Chang, W. Zeng, I. Kamel, and R. Alonso, "Integrated image and speech analysis for content-based video indexing," in *Proc. IEEE Int'l Conf. on Multimedia Computing and Systems*, pp. 306–313, June 1996.

30. T. I. T. Kawashima, K. Tateyama, and Y. Aoki, "Indexing of baseball telecast for content-based video retrieval," in *Proc. IEEE Int'l Conf. on Imag. Proc.*, 1998.

31. Y. Gong, L. T. Sin, C. H. Chuan, H. Zhang, and M. Sakauchi, "Automatic parsing of TV soccer programs," in *IEEE Int'l Conf. on Multimedia Computing and Systems*, pp. 167–174, 1995.

32. M. Yeung, B.-L. Yeo, W. Wolf, and B. Liu, "Video browsing using clustering and scene transitions on compressed sequences," in *Proc. of Multimedia Computing and Networking*. SPIE, vol. 2417, 1995.

33. M. Irani and P. Anandan, "Video indexing based on mosaic representations," *Proceedings of IEEE*, vol. 86, pp. 905–921, May 1998.

34. Y. Li and C.-C. J. Kuo, "A robust video scene extraction approach to movie content abstraction," *International Journal of Imaging Systems and Technology*, vol. 13, no. 5, pp. 236–244, 2004.

35. M. Yeung, B. L. Yeo, and B. Liu, "Extracting story units from long programs for video browsing and navigation," in *Proc. IEEE Conf. on Multimedia Computing and Systems*, 1996.

36. H. J. Zhang, Y. Gong, S. W. Smoliar, and S. Y. Tan, "Automatic parsing of news video," in *Proc. IEEE Conf. on Multimedia Computing and Systems*, pp. 45–54, 1994.

37. P. Aigrain, P. Joly, and V. Longueville, "Medium knowledge based macro segmentation of video into sequences," in *IJCAI Workshop on Intelligent Multimedia Information Retrieval*, pp. 5–14, 1995.

38. H. Aoki, S. Shimotsuji, and O. Hori, "A shot classification method of selecting effective key frames for video browsing," in *Proc. ACM Conf. on Multimedia*, 1995.

39. H. J. Zhang, Y. A. Wang, and Y. Altunbasak, "Content based video retrieval and compression: A unified solution," in *Proc. IEEE Int'l Conf. on Image Processing*, 1997.

40. M. J. Swain and D. H. Ballard, "Color indexing," *Internationl Journal of Computer Vision*, vol. 7, pp. 11–32, 1991.

41. C. A. Bouman, "Cluster: An unsupervised algorithm for modeling Gaussian mixtures," Technical report, School of Electrical Engineering, Purdue University, Oct. 2001. Retrieved fom http://dynamo.ecn.purdue.edu/~bouman/software/cluster.

42. J. Rissanen, "A universal prior for integers and estimation by minimum description length," *Annals of Statistics*, vol. 11, no. 2, pp. 417–431, 1983.

43. Z. Xiong, R. Radhakrishnan, A. Divakaran, and T. S. Huang, "Audio-based highlights extraction from baseball, golf and soccer games in a unified framework," in *Proc. Int'l Conf. on Acoustic, Speech, and Signal Processing*, vol. 5, pp. 628–631, 2003.

44. S. Young, G. Evermann, D. Kershaw, G. Moore, J. Odell et al., *The HTK BOOK VERSION 3.2*, Cambridge: Cambridge University Press, 2003.

45. Z. Xiong, R. Radhakrishnan, A. Divakaran, and T. S. Huang, "Audio-visual sports highlights extraction using coupled hidden Markov models," *Pattern Analysis and Application Journal*, vol. 8, 2005.

46. P. Viola and M. Jones, "Robust real-time object detection," *International Journal of Computer Vision*, vol. 57, no. 2, pp. 137–154, May 2004.

47. A. V. Nefian et al., "A coupled HMM for audio-visual speech recognition," in *Proc. Int'l Conf. on Acoustics, Speech, and Signal Processing*, vol. 2, pp. 2013–2016, 2002.

48. M. Brand, N. Oliver, and A. Pentland, "Coupled hidden Markov models for complex action recognition," in *Proc. Conf. on Computer Vision and Pattern Recognition*, pp. 994–999, June 1997.

49. K. A. Peker, R. Cabasson, and A. Divakaran, "Rapid generation of sports highlights using the mpeg-7 motion activity descriptor," in *Proc. SPIE Conf. on Storage and Retrieval from Media Databases*, vol. 4676, pp. 318–323, 2002.

50. L. R. Rabiner, "A tutorial on hidden Markov models and selected applications in speech recognition," *Proc. IEEE*, vol. 77, no. 2, pp. 257–286, Feb. 1989.

51. Z. Xiong, Y. Rui, R. Radhakrishnan, A. Divakaran, and T. S. Huang, *Handbook of Image & Video Processing*, A Unified Framework for Video Summarization, Browsing, and Retrieval, Academic Press, 2nd ed., 2005.

52. H. V. Jagadish, N. Koudas, and S. Muthukrishnan, "Mining deviants in time series database," in *Proc. 25th Int'l Conf. on Very Large Databases*, pp. 102–113, 1999.

53. S. Muthukrishnan, R. Shah, and J. S. Vitter, "Mining deviants in time series data streams," in *Proc. 16th Int'l Conf. on Scientific and Statistical Database Management*, June 2004.

54. D. Dasgupta and S. Forrest, "Novelty detection in time series using ideas from immunology," in *Proc. Int'l Conf. on Intelligent Systems*, 1999.

55. J. Ma and S. Perkins, "Online novelty detection on temporal sequences," in *Proc. Int'l Conf. on Data Mining*, 2003.

56. C. Shahabi, X. Tian, and W. Zhao, "Tsa-tree: A wavelet based approach to improve efficiency of multi-level surprise and trend queries," in *Proc. Int'l Conf. on Scientific and Statistical Database Management*, 2000.

57. S. Chakrabarti, S. Sarawagi, and B. Dom, "Mining surprising patterns using temporal description length," in *Proc. 24th Int'l Conf. on Very Large Databases*, pp. 606–618, 1998.

58. J. D. Brutlag, "Aberrant behavior detection in time series for network service monitoring," in *Proc. 14th System Administration Conf.*, pp. 139–146, 2000.

59. E. Keogh, S. Lonardi, and W. Chiu, "Finding surprising patterns in time series database in linear time space," in *Proc. 8th ACM SIGKDD Int'l Conf. on Knowledge Discovery and Data Mining*, pp. 550–556, 2002.

60. J. Shi and J. Malik, "Normalized cuts and image segmentation," in *Proc. of IEEE Conf. on Computer Vision and Pattern Recognition*, 1997.

61. R. Perona and W. Freeman, "A factorization approach to grouping," in *Proc. of the 5th European Conf. on Computer Vision*, vol. 1, pp. 655–670, 1998.

62. M. P. Wand and M. C. Jones, *Kernel Smoothing*, London: Chapman & Hall, 1995.

63. S. J. Sheather and M. C. Jones, "A reliable data-based bandwidth selection method for kernel density estimation," *J. R. Statist. Society*, 1991.

64. L. Rabiner and B.-H. Juang, *Fundamentals of Speech Recognition*, 1st ed., Upper Saddle River, NJ: Prentice Hall, 1993.

65. A. Aner and J. R. Kender, "Video summaries through mosaic-based shot and scene clustering," in *Proc. European Conference on Computer Vision*, 2002.

66. R. Radhakrishnan, I. Otsuka, Z. Xiong, and A. Divakaran, "Modelling sports highlights using a time series clustering framework & model interpretation," in *Proc. of SPIE*, 2005.

67. A. Poritz, "Linear predictive hidden Markov models and the speech signal," in *Proc. of ICASSP*, 1982.

68. F. M. Porikli and T. Haga, "Event detection by eigenvector decomposition using object and frame features," in *IEEE Computer Society Conf. on Computer Vision and Pattern Recognition*, June 2004.

69. T. Sikora B. S. Manjunath, P. Salembier, eds., *Introduction to MPEG-7*, Wiley, Hoboken, NJ, 2003.

70. Akutsu et al., "Video indexing using motion vectors," in *Proc. Visual Communications and Image Processing, SPIE*, vol. 1818, pp. 1522–1530, 1992.

71. E. Ardizzone et al., "Video indexing using mpeg motion compensation vectors," in *Proc. IEEE Int'l Conf. on Multimedia Computing and Systems*, 1999.

72. J. W. Davis, "Recognizing movement using motion histograms," Tech. Rep. 487, MIT Media Laboratory Perceptual Computing Section, April 1998.

73. N. Dimitrova and F. Golshani, "Motion recovery for video content analysis," *ACM Trans. Information Systems*, vol. 13, no. 4, pp. 408–439, Oct. 1995.

74. K.-I. Lin, V. Kobla, D. Doermann, and C. Faloutsous, "Compressed domain video indexing techniques using dct and motion vector information in mpeg video," in *Proc. SPIE Conf. on Storage and Retrieval for Image and Video Databases V*, vol. 3022, pp. 200–211, 1997.

75. E. Sahouria, "Video indexing based on object motion," M.S. thesis, UC Berkeley, Dept. of EECS, 1999.

76. S. R. Kulkarni, Y. P. Tan, and P. J. Ramadge, "Rapid estimation of camera motion from compressed video with application to video annotation," *IEEE Trans. on Circuits and Systems for Video Technology*, vol. 10, no. 1, pp. 133–146, Feb. 2000.

77. Y. T. Tse and R. L. Baker, "Camera zoom/pan estimation and compensation for video compression," in *Proc. SPIE Conf. on Image Processing Algorithms and Techniques II*, pp. 468–479, 1991.

78. W. Wolf, "Key frame selection by motion analysis," in *Proc. ICASSP*, pp. 1228–1231, 1996.

79. H. J. Meng, H. Sundaram, D. Zhong, S. F. Chang, and W. Chen, "A fully automated content-based video search engine supporting multi-objects spatio-temporal queries," *IEEE Transactions on Circuit and Systems for Video Technology*, vol. 8, no. 5, pp. 602–615, Sep. 1998.

80. D. Petkovic, W. Niblack, D. B. Ponceleon, "Updates to the qbic system," in *Proc. IS&T SPIE, Storage and Retrieval for Image and Video Databases VI*, vol. 3312, pp. 150–161, 1996.

81. URL, http://www.virage.com, *Virage system*.

82. URL, http://www.qbic.almaden.ibm.com, *QBIC system*.

83. URL, http://www.ctr.columbia.edu/videoq, *VideoQ system*.

84. T. Kaneko and O. Hori, "Results of spatio-temporal region ds core/validation experiment," in *ISO/IEC JTC1/SC29/WG11/MPEG99/M5414*, Dec. 1999.

85. B. Mory and S. Jeannin, "Video motion representation for improved content access," *IEEE Transactions on Consumers Electronics*, Aug. 2000.

86. K. Asai, A. Divakaran, A. Vetro, and H. Nishikawa, "Video browsing system based on compressed domain feature extraction," *IEEE Trans. Consumer Electronics*, Aug. 2000.

87. P. Salembier and J. Llach, "Analysis of video sequences: Table of contents and index creation," in *Proc. Int'l Workshop on very low bit-rate video coding (VLBV'99)*, pp. 52–56, Oct. 1999.

88. "Mpeg-7 visual part of the xm 4.0," in *ISO/IEC MPEG99/W3068*, Dec. 1999.

89. S. Jeannin and A. Divakaran, "Mpeg-7 visual motion descriptors," *IEEE Trans. on Circuits and Systems for Video Technology*, special issue on MPEG-7, vol. 11, no. 6, June 2001.

90. A. Divakaran and K. Peker, "Video summarization using motion descriptors," in *Proc. SPIE Conf. on Storage and Retrieval from Multimedia Databases*, Jan. 2001.

91. A. Divakaran, K. Peker, and H. Sun, "Constant pace skimming and temporal subsampling of video using motion activity," in *Proc. ICIP*, 2001.

92. Y. Rui, T. S. Huang, M. Ortega, and S. Mehrotra, "Relevance feedback: A power tool for interactive content-based image retrieval," *IEEE Trans. on Circuits and Systems for Video Technology*, special issue on segmentation, description, and retrieval of video content, vol. 8, pp. 644–655, Sep. 1998.

93. S. Tong and E. Y. Chang, "Support vector machine active learning for image retrieval," in *ACM Int'l Conf. on Multimedia*, pp. 107–118, Oct. 2001.

94. M. Naphade and T. S. Huang, "A probabilistic framework for semantic video indexing, filtering and retrieval," *IEEE Transactions on Multimedia*, vol. 3, no. 1, pp. 141–151, March 2001.

95. M. Casey, "Mpeg-7 sound-recognition tools," *IEEE Transactions on Circuits and Systems for Video Technology*, vol. 11, no. 6, pp. 737–747, June 2001.

96. J. R. Smith, M. R. Naphade, and A. Natsev, "Multimedia semantic indexing using model vectors," in *Proc. IEEE Int'l Conf. on Multimedia & Expo (ICME)*, July 2003.

97. L. Chaisorn, T.-S. Chua, C.-H. Lee, and Q. Tian, "Hierarchical approach to story segmentation of large broadcast news video corpus," in *Proc. IEEE Int'l Conf. on Multimedia & Expo (ICME)*, June 2004.

98. Y. Rui, T. S. Huang, and S. Mehrotra, "Constructing table-of-content for videos," *Journal of Multimedia Sys.*, vol. 7, pp. 359–368, Sep. 1999.

99. C.-W. Ngo, A. Velivelli, and T. S. Huang, "Detection of documentary scene changes using audio-visual fusion," in *Int'l Conf. on Image and Video Retrieval*, July 2003.

100. H. Sundaram and S. Chang, "Audio scene segmentation using multiple models, features and time scales," in *IEEE Int'l Conf. on Acoustics, Speech, and Signal Processing*, Istanbul, Turkey, June 2000.

101. C.-W. Ngo, *Analysis of spatio-temporal slices for video content representation*, Ph.D. dissertation, Hong Kong University of Science and Technology, 2000.

102. Y. Rui, T. S. Huang, and S. Mehrotra, "Content-based image retrieval with relevance feedback in MARS," in *Proc. IEEE Int'l Conf. on Image Proc.*, 1997.

103. J. A. Schmidt, "Object and camera parameter estimation using mpeg motion vectors," M.S. thesis, University of Illinois at Urbana-Champaign, 1998.

104. Y. Rui, T. S. Huang, and S.-F. Chang, "Image retrieval: Current techniques, promising directions and open issues," *Journal of Visual Communication and Image Representation*, vol. 10-4, pp. 39–62, 1999.

105. A. Kudhinsky, C. Pering, M. L. Creech, D. Freeze, B. Serra, and J. Gvvizdka, "FotoFile: A consumer multimedia organization and retrieval system," in *Proceedings of CHI'99*, pp. 496–503, 1999.

106. H. D. Wactlar, T. Kanadeand, M. A. Smith, and S. M. Stevens, "Intelligence access to digital video: Informedia project," *IEEE Computer*, vol. 29-5, pp. 46–52, 1996.

107. H. Rowley, S. Baluja, and T. Kanade, "Neural network-based face detection," *IEEE Patt. Anal. Mach. Intell.*, vol. 20, pp. 22–38, 1998.

108. ACLU, http://archive.aclu.org/features/f110101a.html, 2000.

109. S. Russell and P. Norvig, *Artificial Intelligence: A Modern Approach*, 2nd ed., Upper Saddle River, NJ: Prentice Hall, 1995.

110. H. Berliner, "The B* tree-search algorithm: A best-first proof procedure," *Artificial Intelligence*, vol. 12, no. 1, pp. 23–40, 1979.

111. B. Frey, *Graphical Models for Machine Learning and Digital Communication*, Cambridge: MIT Press, 1998.

About the Authors

Ziyou Xiong is a senior research engineer/scientist at the Dynamic Modelling and Analysis group of the United Technologies Research Center in East Hartford, Connecticut. He received his bachelor's degree from Wuhan University, Hubei Province, China, in July 1997. He received his master's degree in electrical and computer engineering from the University of Wisconsin in Madison in December 1999. He received his Ph.D. in electrical and computer engineering from the University of Illinois at Urbana-Champaign (UIUC). He was also a research assistant with the Image Formation and Processing Group of the Beckman Institute for Advanced Science and Technology at UIUC from January 2000 to August 2004. In the summers of 2003 and 2004, he worked on sports audio-visual analysis at Mitsubishi Electric Research Labs in Cambridge, Massachusetts. His current research interests include image and video analysis, video surveillance, computational audio-visual scene analysis, pattern recognition, machine learning, and related applications. He has published several journal and conference papers and has written invited book chapters on audio-visual person recognition, image retrieval, and sports video indexing, retrieval, and highlight extraction. He has also coauthored a book titled *Facial Analysis from Continuous Video with Application to Human-Computer Interface* (Kluwer, 2004).

    Regunathan Radhakrishnan received his B.E. (with honors) in electrical engineering and M.Sc. (with honors) in chemistry from Birla Institute of Technology and Science (BITS), Pilani, India, in 1999. He worked as a Digital Signal Processing (DSP) engineer in Multimedia Codecs Group at SASKEN Communication Technologies Ltd. in Bangalore, India from 1999–2000. He received his M.S. and Ph.D. degrees in electrical engineering from Polytechnic University, Brooklyn, New York, in 2002 and 2004, respectively. He was a research fellow in the Electrical and Computer Engineering (ECE) department and an intern at Mitsubishi Electric Research Labs in Cambridge, Massachusetts, during his graduate studies. He joined Mitsubishi Electric Research Laboratories (MERL) in 2005 as a visiting researcher. His current research interests include audio classification, multimedia mining, digital watermarking, and content security and data mining. He has published several conference papers, as well as five journal papers and three book chapters on multimedia content analysis and security.

Ajay Divakaran received his B.E. (with honors) in electronics and communication engineering from the University of Jodhpur, India, in 1985, and his M.S. and Ph.D. degrees from Rensselaer Polytechnic Institute, Troy, New York, in 1988 and 1993, respectively. He was an assistant professor with the Electrical and Computer Engineering (ECE) Department, University of Jodhpur, in 1985–1986. He was a research associate at the Electrical and Computer Engineering (ECE) Department, Indian Institute of Science, in Bangalore in 1994–1995. He was a scientist with Iterated Systems, Inc., Atlanta, Georgia, from 1995 to 1998. He joined Mitsubishi Electric Research Laboratories (MERL) in 1998 and is now a senior team leader and senior principal member of the technical staff. He has been a key contributor to the MPEG-7 video standard. His current research interests include video and audio analysis, summarization, indexing and compression, and related applications. He has published several journal and conference papers, as well as six invited book chapters on video indexing and summarization. He has supervised three doctoral theses. He currently serves on program committees of key conferences in the area of multimedia content analysis. He currently leads the Data and Sensor Systems Team at the Technology Laboratory of MERL.

Yong Rui is a researcher in the Communication and Collaboration Systems (CCS) Group in Microsoft Research, where he leads the Multimedia Collaboration Team. Dr. Rui is a senior member of the Institute of Electrical and Electronics Engineers (IEEE) and a member of the Association of Computing Machinery (ACM). He is on the editorial board of International Journal of Multimedia Tools and Applications. He received his Ph.D. from the University of Illinois at Urbana-Champaign (UIUC).

Dr. Rui's research interests include computer vision, signal processing, machine learning, and their applications in communication, collaboration, and multimedia systems. He has published one book (*Exploration of Visual Data*, Kluwer Academic Publishers), six book chapters, and more than 60 referred journal and conference papers in these areas. Dr. Rui was on program committees of ACM Multimedia, IEEE Computer Vision and Pattern Recognition (CVPR), IEEE European Conference on Computer Vision (ECCV), IEEE Asian Conference on Computer Vision (ACCV), IEEE International Conference on Image Processing (ICIP), IEEE International

Conference on Acoustics, Speech, and Signal Processing (ICASSP), IEEE International Conference on Multimedia Expo (ICME), Society of Photo-optical Instrumentation Engineers (SPIE), Information Technologies and Communications (ITCom), International Conference on Pattern Recognition (ICPR), and Conference on Image and Video Retrieval (CIVR), among others. He was a co-chair of IEEE International Workshop on Multimedia Technologies in E-Learning and Collaboration (WOMTEC) 2003, the demo chair of ACM Multimedia 2003, and a co-tutorial chair of ACM Multimedia 2004. He was on the National Science Foundation (NSF) review panel and National Academy of Engineering's Symposium on Frontiers of Engineering for outstanding researchers.

Thomas S. Huang received his B.S. degree in electrical engineering from National Taiwan University, Taipei, Taiwan, China; and his M.S. and Sc.D. degrees in electrical engineering from the Massachusetts Institute of Technology (MIT), Cambridge, Massachusetts. He was on the faculty of the Department of Electrical Engineering at MIT from 1963 to 1973, and he was on the faculty of the School of Electrical Engineering and director of its Laboratory for Information and Signal Processing at Purdue University from 1973 to 1980. In 1980, he joined the University of Illinois at Urbana-Champaign, where he is now William L. Everitt Distinguished Professor of Electrical and Computer Engineering and research professor at the Coordinated Science Laboratory. He also serves as head of the Image Formation and Processing Group at the Beckman Institute for Advanced Science and Technology and co-chair of the institute's major research theme, human computer intelligent interaction.

During his sabbatical leaves, Dr. Huang has worked at the MIT Lincoln Laboratory, the IBM Thomas J. Watson Research Center, and the Rheinishes Landes Museum in Bonn, West Germany, and he has held visiting professor positions at the Swiss Institutes of Technology in Zurich and Lausanne, the University of Hannover in West Germany, INRS-Telecommunications of the University of Quebec in Montreal, Canada, and the University of Tokyo, Japan. He has served as a consultant to numerous industrial firms and government agencies, both in the United States and abroad.

Dr. Huang's professional interests lie in the broad area of information technology, especially the transmission and processing of multidimensional signals. He has published 14 books and more than 500 papers in network theory, digital filtering, image processing, and computer vision. He is a member of the National Academy of Engineering; a foreign member of the Chinese Academies of Engineering and Sciences; and a fellow of the International Association of Pattern Recognition, IEEE, and the Optical Society of America. He has received a Guggenheim Fellowship, an A.V. Humboldt Foundation Senior U.S. Scientist Award, and a fellowship from the Japan Association for the Promotion of Science. He received the IEEE Signal Processing Society's Technical Achievement Award in 1987 and the Society Award in 1991. He was awarded the IEEE Third Millennium Medal in 2000. Also in 2000, he received the Honda Lifetime Achievement Award for "contributions to motion analysis." In 2001, he received the IEEE Jack S. Kilby Medal. In 2002, he received the King-Sun Fu Prize, International Association of Pattern Recognition; and the Pan Wen-Yuan Outstanding Research Award. He is a founding editor of the *International Journal of Computer Vision, Graphics, and Image Processing* and editor of the *Springer Series in Information Sciences*, published by Springer Verlag.

Dr. Huang initiated the first International Picture Coding Symposium in 1969 and the first International Workshop on Very Low Bitrate Video Coding in 1993. Both meetings have become regular events (held every 12 to 18 months) and have contributed to the research and international standardization of image and video compression. He also (together with Peter Stucki and Sandy Pentland) initiated the International Conference on Automatic Face and Gesture Recognition in 1995. This conference has also become a regular event and provides a forum for researchers in this important and popular field.

Index